17th International Scientific Conference - Civil Engineering and Building Services

IOP Conference Series: Materials Science and Engineering
Volume 1283

Brasov, Romania
3-4 November 2022

Editors:

Ioan Tuns
Radu Muntean
Dorin Radu
Teofil-Florin Gălățanu
Christiana Emilia Cazacu

ISBN: 978-1-7138-7496-6
ISSN: 1757-8981

Printed from e-media with permission by:

Curran Associates, Inc.
57 Morehouse Lane
Red Hook, NY 12571

Some format issues inherent in the e-media version may also appear in this print version.

Institute of Physics
Dirac House, Temple Back
Bristol BS1 6BE UK

Phone: 44 1 17 929 7481
Fax: 44 1 17 920 0979

techtracking@iop.org

Additional copies of this publication are available from:

Curran Associates, Inc.
57 Morehouse Lane
Red Hook, NY 12571 USA
Phone: 845-758-0400
Fax: 845-758-2633
Email: curran@proceedings.com
Web: www.proceedings.com

TABLE OF CONTENTS

International Conference CIBv2022 Civil Engineering and Building Services

Peer Review Statement

Experimental Investigation of Parameters Influencing Mechanical Properties of Hemp Concretes Using Design of Experiment Method 1
L Adam, D N Isopescu, D Lepadatu

A Non-Destructive Method for Fixing Placards with Masonry Structures 19
L Bianco, A Bianco

Perlite Concrete: A Review 28
A Cojocaru, D N Isopescu, S G Maxineasa

Heavy and Light Masonry Walls Impact on Medium Height Reinforced Concrete Frames Buildings 37
S Constantinescu

Technology for Utilization and Neutralization of Municipal Waste 49
V Ezhov, V Pakhomov, D Fedyanin, I Saikov

Studies on Heavy Duty Pavements Reinforced with Steel Fibers for Port Infrastructures 55
P Ionițoaiei, A Rotaru

Eco-Concrete for Obtaining "Green" Construction Elements 63
B I Luca, A Panțiru, A Timu, M Bărbuță, L I Diaconu, M Rujanu, A C Diaconu

Flood Hazard Assessment in Montenegro Through the Application of Two-Dimensional Hydraulic Modelling 71
M Ostojić, I Ćipranić, M Jevrić, G Sekulić, A Medojević

Analysis of Mortars with Old Plastering Waste Using State-Of-The-Art Method: Nuclear Magnetic Resonance 79
C Saitis, D L Manea, R Fechete, L M Plesa, L Saitis

Laboratory Study Regarding Recipes and Strengths of Concrete Produced with Reactive Powders and Metallic Fibers 89
S Scripcă, M Bărbuță, V Panaite

Steel Hall Constructive Solution Beam with a Full Web, Optimization of Sections and Joints 100
F Tepes-Onea, N Janczak

Identification Method of Unbalanced Fault Characteristics of Horizontal Shaft of Rotary Table of Drilling Rig 106
Z Wang, C Q Lu, J Ma, S Y Du

Thermal Characteristics Analysis of Bearing in Grinding System of Grinding Robot 113
Z Wang, Z C Yin, Z N Wang, C M Chen

Reuse of Mixed Masonry Wastes in Plastering Mortars 121
A Zaharie, M L Țințișan, AC Siomin, D L Manea, M L Pleșa

Author Index

International Conference CIBv2022
Civil Engineering and Building Services

3rd - 4th November 2022, Brașov, Romania

Selected, peer reviewed papers from the International Conference Civil Engineering and Building Services CIBv2022

Edited by:

Ioan TUNS
Transilvania University of Brasov, Romania

Radu MUNTEAN
Transilvania University of Brasov, Romania

Dorin RADU
Transilvania University of Brasov, Romania

Teofil-Florin GĂLĂȚANU
Transilvania University of Brasov, Romania

Christiana Emilia CAZACU
Transilvania University of Brasov, Romania

IOP Conf. Series: Materials Science and Engineering 1283 (2023) 011001 doi:10.1088/1757-899X/1283/1/011001

FOREWORD

The International Scientific Conference CIBv represents one of the most important and traditionally scientific events of Civil Engineering and Building Services in Romania, enjoying increasingly higher visibility at the international level.

The aim of the CIBv conference is to provide a platform for researchers, engineers, academicians as well as professionals from all over the world to present their research results and development activities. It also provides opportunities for the participants to exchange new ideas and application experiences, to establish business or research relations and to find other partners and even friends for future collaboration.

The Conference has been organized each year at Transilvania University of Brașov, since 2000. In November 2022, Faculty of Civil Engineering was honored to organize, in partnership with Shenyang Jianzhu University from China and to host the 17th edition of "International Scientific Conference – Civil Engineering and Building Services CIBv2022" in Brașov. More than 110 participants from China and Romania, as well as from United States of America, Serbia, Japan, Germany, Russia and Greece submitted over 62 papers to the scientific board which were presented during the conference, which was organized on site. After a peer-review process (which included for each article at least two reviewers), 17 papers were finally accepted for publication in the CIBv2022 proceedings volume.

We would like to address special thanks to all those who were involved in the success achieved by the CIBv2022 International Conference, as well as to our distinguished publisher.

Prof. Eng. Ioan TUNS, PhD

Dean of Faculty of Civil Engineering,
Transilvania University of Brașov, Romania

IOP Conf. Series: Materials Science and Engineering 1283 (2023) 011001 doi:10.1088/1757-899X/1283/1/011001

International Conference CIBv2022
Civil Engineering and Building Services

Organized by
Faculty of Civil Engineering, Transilvania University of Brasov, Romania in collaboration with **Shenyang Jianzhu University, China** and **Technical University of Moldova, Faculty of Constructions, Geodesy and Cadastre**

Honorary Committee

Ioan Vasile ABRUDAN, Rector of Transilvania University Braşov

Weidong YAN, President of Shenyang Jianzhu University

Yupeng LI, Vice president of Shenyang Jianzhu University

Viorel BOSTAN, Rector of the Technical University of Moldova

Dan STEMATIU, President of Civil Engineering and Urbanism section of the Romanian Academy of Technical Sciences Professor of Technical University of Civil Engineering of Bucharest

International academic scientific committee
Radu BĂNCILĂ, Politehnica University Timişoara, Romania
Miroslav BEŠEVIĆ, University of Novi Sad, Serbia
Alina BĂRBULESCU, Technical University C.E. Bucureşti, Romania
Marius Florin BOTIŞ, Transilvania University of Braşov, Romania
Andrei BURLACU, Technical University „Gh. Asachi" Iaşi, Romania
Natalia BREDIKHINA , Southwest State University, Kursk Russia
Ján BUJŇÁK, Technical University Žilina, Slovakia
Sorin BURCHIU, Technical University C.E. Bucureşti, Romania
Mircea Ion BUZDUGAN, Technical University Cluj-Napoca, Romania
Camelia CERBU, Transilvania University of Braşov, Romania
Nicoleta COBÂRZAN, Technical University Cluj-Napoca, Romania
Ioan Lucian CÎRSTOLOVEAN, Transilvania University of Braşov, Romania
Nicolae CHIRA, Technical University Cluj-Napoca, Romania
Vasilica CIOCAN, Technical University „Gh. Asachi" Iaşi, Romania
Ion CREȚU, Technical University of Moldova
Mircea Radu DAMIAN, Technical University C.E. Bucureşti, Romania
Sorin DAN, Universitatea Politehnica Timişoara, Romania
Liviu Geo DRUGHEAN, Technical University C.E. Bucureşti, Romania
Liviu DUMITRESCU, AIIR, Romania
Alexandru Cătălin FILIP, Transilvania University of Braşov, Romania
Wolfgang FRANCKE, University of Applied Sciences Konstanz, Germania
Radomir FOLIĆ, University of Novi Sad, Serbia
Teofil GĂLĂȚANU, Transilvania University of Braşov, Romania
Marijana HADZIMA-NYARKO, Josip Juraj Strossmayer University of Osijek, Croatia
Sorin HERBAN, University Politehnica Timişoara, Romania
Kailiang HUANG, Shenyang Jianzhu University, China
Nicolae Fani IORDAN, Transilvania University of Braşov, Romania
Danijel KUKARAS, University of Novi Sad, Serbia

IOP Conf. Series: Materials Science and Engineering 1283 (2023) 011001 doi:10.1088/1757-899X/1283/1/011001

Organizing committee

Peer Review Statement

All papers published in this volume have been reviewed through processes administered by the Editors. Reviews were conducted by expert referees to the professional and scientific standards expected of a proceedings journal published by IOP Publishing.

- **Type of peer review:** Single Anonymous
- **Conference submission management system:** Morressier
- **Number of submissions received:** 32
- **Number of submissions sent for review:** 25
- **Number of submissions accepted:** 17
- **Acceptance Rate (Submissions Accepted / Submissions Received × 100):** 53.1
- **Average number of reviews per paper:** 2
- **Total number of reviewers involved:** 50
- **Contact person for queries:**
  Name: Radu Muntean
  Email: radu.m@unitbv.ro
  Affiliation: Transilvania University of Brasov

Experimental investigation of parameters influencing mechanical properties of hemp concretes using design of experiment method

L Adam[1*], D N Isopescu[1] and D Lepadatu[1]

[1] "Gheorghe Asachi" Technical University of Iasi-Romania, Faculty of Civil Engineering and Building Services, Iasi, Romania

* laurentiu.adam@student.tuiasi.ro

Abstract. Global concerns about climate change have led to the development of building materials that use plant biomass with the added benefit of sequestering carbon from the atmosphere. Lightweight concrete made with hemp, is a material with huge potential in terms of its ability to achieve hygro-thermal performances close to those of polystyrene. This ecological material, referred to in the literature as hemp concrete, is made of some components that, depending on the chosen combination, can generate different results of the physic- mechanical properties. The method is based on a series of experiments that are generated using a program called Design of Experiments (DoE). This program uses statistical analysis of a series of experiments performed by varying one of the input parameters, like the amount of hydrated lime or cement. Following this analysis, it is possible to identify those elements used in the composition, which produced a significant influence on the physical-mechanical properties of hemp concrete. Moreover, this method also allows a quantification of the influences so that the parameters can be ranked, with a minimum number of experiments. The method is, in the same time, a modern and an economical technique used for experimental planning. The aim of the research is to quantify numerically the influence of each component element of the mixture that determines variations of the physical-mechanical properties in hemp concrete. This study will contribute in the future to the optimization of compositions, which will be differentiated on the basis of mechanical performance or thermal insulation criteria.

1. Introduction

Hemp concrete is a composite material, for the production of which are used hemp woodchips, a mix of mineral binder based on hydrated lime and water. Even the first use of this vegetable mixture with mineral binders took place in 1986 [1], the interest in this natural composite grows with the concern of scientists towards to discovery of solutions that reduce the generation of carbon dioxide or finding ways to absorb the greenhouse gas from the atmosphere to improve global climate change. This material has low load-bearing properties, compressive strength reaching values below 1 MPa [2-4], but hemp concrete is currently being analyzed due to the performances that it records as a sound-absorbing material, thermal insulator and its ability to regulate the hygro-thermal parameters of the indoor air. Thus, having low heat transfer coefficient, being capable of absorbing up to four times the dry mass of the hemp woodchips and managing to attenuate sound vibrations [5-10], it qualifies on the list of sustainable materials as an ally in the fight against climate warming and a solution to the energy crisis in the field of construction, being at the same time an ecological material that also solves the need to maintain a healthy and comfortable interior environment in living spaces [11-17]. The need to obtain economic and ecological performing materials requires conducting experiments to determine optimized mixtures specialized on thermo-mechanical properties [18-20]. Optimization offers, in addition to the possibility of projecting performances, the advantage of saving resources as long as precise quantities were used in the mixture for the desired effects. In order to achieve this optimization, the analysis of the influence of each variable component introduced in the mixture must be carried out.

In the present work the elements of the mixture will be analyzed from the point of view of their influence on the properties of hemp concrete. The mass variation of hydrated lime, cement, water and sodium silicate [21] is the criteria that will determine obtaining different mixtures. This variation is given by the complete factorial plan obtained based on the minimum and maximum quantities of the ingredients used in the mixture. Sixteen different hemp concrete mixtures will be generated using the design of experiment method - DoE [22-26] – that allows, by performing the smallest possible number of tests, to obtain the maximum possible response of information that can characterize the cumulative influence of the components in the mixture on the physical-mechanical properties of the material obtained.

The values will be used to determine the ranking of the influence degree regarding the interaction between the ingredients used and the hemp powder mass included in the mixture. The goal is to understand how the different combinations of binders, water and sodium silicate produce variations in the mechanical behavior and the variation of the heat transfer coefficient of the material made by analyzing the cubic samples. DoE is the advanced method of designing experiments that uses statistical calculation in fields such as biotechnology and industrial engineering, but this method is also applied in product marketing. The method is based on the principle that for each independent input variable there is a variable response of a parameter whose magnitude is dependent on the input parameter. Thus, a prediction of the result can be done depending on the size of the initial conditions.

Another advantage is that by using this method it is possible to identify those control variables that must be kept constant so that the results are not disturbed/altered by auxiliary factors such as mixing time and speed, relative humidity and temperature of the environment, the humidity of the hemp woodchips, the time of introducing the ingredients in the mixture, the degree of grinding of the binders used, the size and shape of the hemp woodchips, the use of additives or the quality and temperature of the water used to hydrate the binders are carried out.

The DoE method allows the calculation of the deviation between the output results from the statistical calculation and the laboratory measurements, and makes possible to get information regarding the conditions in which the experiment is carried out and the accuracy of the operator's execution. The method is, at the same time, a tool for verifying compliance with the technology of making the mixture, correct measurements of the quantities in the mixtures and carrying out the tests on the samples under the same conditions.

All these variables for hemp concrete can be constituted in a cause-effect diagram, (Ishikawa diagram) shown in figure 1. From the multitude of variables identified, four variables were chosen for this study. The identification, quantification and ranking of the independent variables (factors), considered as input quantities, on the physical-mechanical properties, will be displayed graphically according to the Pareto principle.

Variability analysis, presented through ANOVA tables, [27], will numerically display the variable parameters, respectively the combined influence of the response properties measured in the laboratory tests. Thus, the researcher has a modern experimental planning tool and, at the same time, a powerful analysis and optimization of generally non-linear processes, in which it is very difficult to accurately assess the causes that determined the changes of the characteristic responses, in the process studied.

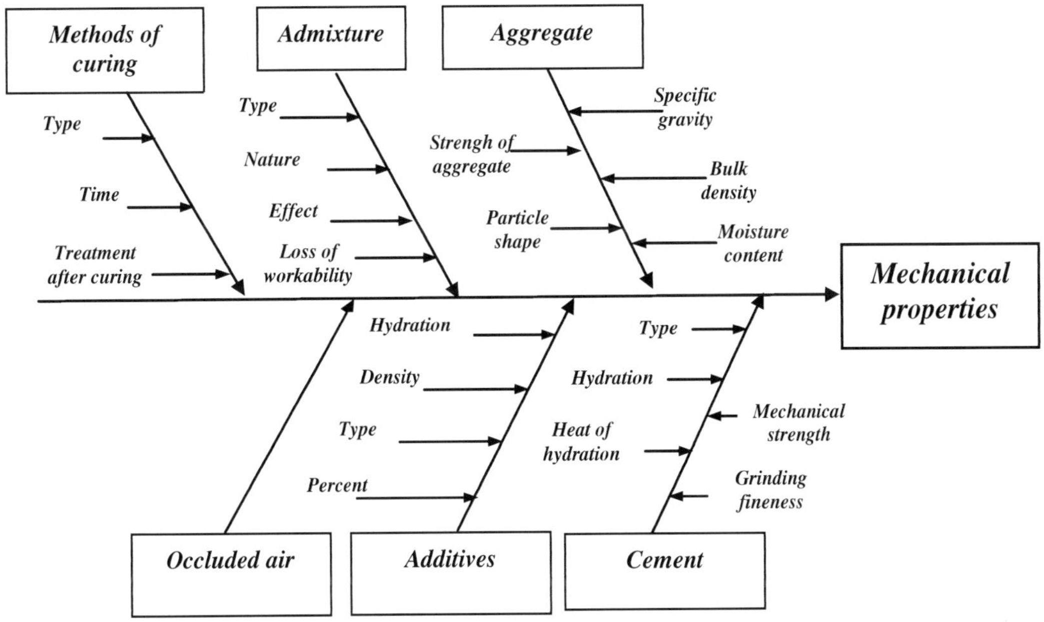

Figure 1. Ishikawa diagram of hemp concrete.

2. Materials

2.1. Preparing the base material

The materials used for making the mixture where hemp shiv, hydrated lime, Portland cement, sodium silicate and water. From a volumetric point of view, hemp powder is the basic material in the hemp concrete mixture. After harvesting, the hemp stalks were chopped by an electrical device consisting of a support table and cutting knives helically mounted on a drum driven, and by a spindle and a flywheel mounted on the spindle, used for separate the fiber from the woody part of the plant, figure 2. The hemp woodchips obtained were uniformed by making two passes of the material through the "hammer mill" device shown in figure 3. At the end of the decortication and chopping processes, from the hemp shiv the possible earth particles or stones are eliminated.

The hemp woodchips are shown in figure 4 and in the table 1, the particle size distribution is detailed.

Figure 2. Electrical device used for decortication of hemp stems.

Figure 3. Chopping hemp steam core with a hammer mill device.

Figure 4. Particle size distribution.

In table 1 is presented the distribution of particle size fractions relative to the total weight of the hemp shiv obtained.

Table 1. Distribution of particle size fractions.

Type of particle size	Dimension [mm]	Weight kg	Bulk density kg/m^3	Mass percent %
F1	0 - 4	0.579	193	8.92
F2	5 - 10	1.540	110	23.73
F3	11 - 35	4.371	93	67.35
Total	0 - 35	6.490	101.4	100.00

2.2. Mineral binders used for the mix

Hydrated lime and Portland cement were used to obtain the binder mix in mass proportions according to the factorial plan based on DoE. These proportions are detailed in the section that includes the method used to plan the experiment and the mixtures, according to which, the cubic samples were casted. Some specific details of the mineral binders used in mixture are presented as following: the cement characteristics in table 2 and the hydrated lime components in table 3.

Table 2. Characteristics of cement.

Characteristics	U.M.	Values
Clinker (K) content	%	65÷79
Furnace slag (S) and chalk (LL) content	%	21÷35
Sulphate content in the form of SO_3	%	≤ 3.5
Standard compression strength	MPa	42.5÷62.5
Expansion stability	mm	≤ 3

Table 3. Components of hydrated lime.

Components	U.M.	Values
CaO + MgO	%	≥ 80
Free lime Ca $(OH)_2$	%	≥ 65
MgO	%	≤ 5
CO_2	%	≤ 15
SO_3	%	≤ 2
Free water	%	≤ 2
Air content	%	≤ 12

2.3. Sodium silicate solution

Sodium silicate or glass water was used as an additive to increase the binding speed of the mineral binder matrix due to the reduction of the water mass required for cement hydration. Sodium silicate liquid solution was used with a concentration of 35-40%. The additive was added to the mixing water in the proportion of 5% by mass of water.

3. Method

The method used for experimental planning is Design of Experiment – DoE. It is an advanced method of experimental research planning that allows the investigation of several variable parameters at the same time. This modern method replaces the classic methods, and reduces the number of laboratory tests encountered in researches in which there are several parameters that can vary simultaneously.

The steps of the research presented in case study, according to DoE method, are the following:

1. Define objective

The objective is to investigate which are the parameters that influence the thermo-mechanical properties of hemp concrete, namely the identification of those variable factors that can influence the heat transfer coefficient.

2. The choice of independent variables

The hemp concrete mixture is made of hemp woodchips, Portland cement, hydraulic lime, water and sodium silicate. The mixture components, presented in table 4, are the independent variables.

Table 4. Independent variables.

Independent variables	Quantity (minimum), [g]	Quantity (maximum), [g]
Hydraulic lime	492	714
Portland Cement	165	510
Water	657	1224
Sodium silicate	33	63

3. The choice of dependent variables (Outputs variables or system responses)

The dependent variables are the response of the system. In the presented research, the compression resistance and the heat transfer coefficient are the dependent variables.

4. Selection of plan type and method of analysis

The analysis is based on the complete factorial plan obtained by combining mixtures with different values for the independent variables chosen on point 2. In table 5, according to the factorial plan, the hemp concrete mixtures are presented.

Table 5. Full factorial plan 2^n (n=4).

Mixture	Mass of the independent variables			
	Hydraulic lime [g]	Portland cement [g]	Water [g]	Sodium Silicate [g]
1	492	165	657	33
2	714	165	657	33
3	492	510	657	33
4	714	510	657	33
5	492	165	1224	33
6	714	165	1224	33
7	492	510	1224	33
8	714	510	1224	33
9	492	165	657	63
10	714	165	657	63
11	492	510	657	63
12	714	510	657	63
13	492	165	1224	63
14	714	165	1224	63
15	492	510	1224	63
16	714	510	1224	63

5. Perform the experimental design

Carrying out the experiment first involves making the samples. The amount of hemp, figure 4, were mixed using only the F2 and F3 particle size fraction, (table 1). Then the hemp woodchips were mixed 1:1 with a quantity of water equal with their dry mass. The technology used for making the mixture was as follows:

- add the amount of moistened hemp powder;
- premix the hydrated lime with Portland cement according to the quantities specified in mixture;
- premix the sodium silicate with water used to hydrate the binder mix;
- add the binder mix to the hemp shiv;
- mix until obtain homogeneous;
- add the water solution with sodium silicate and homogenizing;
- cast fresh hemp concrete in a cubic shape with a side size of 15 cm, figures 5 and figure 6.

The casting of hemp concrete samples was carried out according to EN 12390-1 [28].

Figure 5. Casting the cubic samples.

Figure 6. All the cubic samples after 28 days of curing.

6. Statistical analysis of the obtained results

The last step is the statistical analysis. In subchapter 4.5, the results obtained in this study are analyzed.

4. Results and discussion

4.1. Mass loss after 28 days of curing time

The variation of the samples mass is present in table 6, according to the mass measurements schedule.

Table 6. Mass variation during the 28 days of curing time.

Mixture	Day 1	2	5	8	12	15	19	22	25	27	28
	Mass values, [g]										
1	2538	2503	2338	2149	1981	1912	1834	1798	1763	1740	1739
2	2788	2758	2611	2414	2248	2167	2090	2052	2012	1992	1991
3	2978	2952	2821	2650	2520	2463	2415	2378	2349	2333	2330
4	2859	2846	2736	2593	2467	2418	2372	2342	2318	2304	2302
5	2798	2772	2624	2405	2130	1981	1829	1770	1724	1692	1686

6	3176	3149	3015	2807	2565	2421	2258	2185	2135	2097	2091
7	3189	3163	3026	2846	2628	2526	2414	2360	2333	2307	2303
8	3564	3536	3390	3207	2932	2778	2627	2552	2493	2452	2443
9	2900	2875	2728	2551	2310	2204	2109	2044	1996	1969	1960
10	2984	2962	2825	2644	2425	2334	2240	2181	2139	2112	2108
11	3000	2974	2863	2689	2531	2472	2407	2377	2344	2319	2318
12	3025	3000	2880	2723	2598	2530	2465	2432	2393	2371	2369
13	3302	3274	3108	2873	2563	2380	2174	2094	2016	1956	1951
14	3458	3430	3258	3023	2706	2548	2338	2249	2161	2112	2104
15	3616	3585	3393	3155	2849	2716	2579	2516	2457	2425	2422
16	3775	3738	3524	3262	2951	2828	2708	2655	2607	2571	2568

The maximum loss of mass was obtained for sample 13 and sample 5 as it shown in figure 7. These two cubic samples were made using the highest amount of water and the lowest amount of Portland cement. The lowest loss of mass was obtained by samples 4 and 12, which have the lowest amount of water and the highest amount of Portland cement.

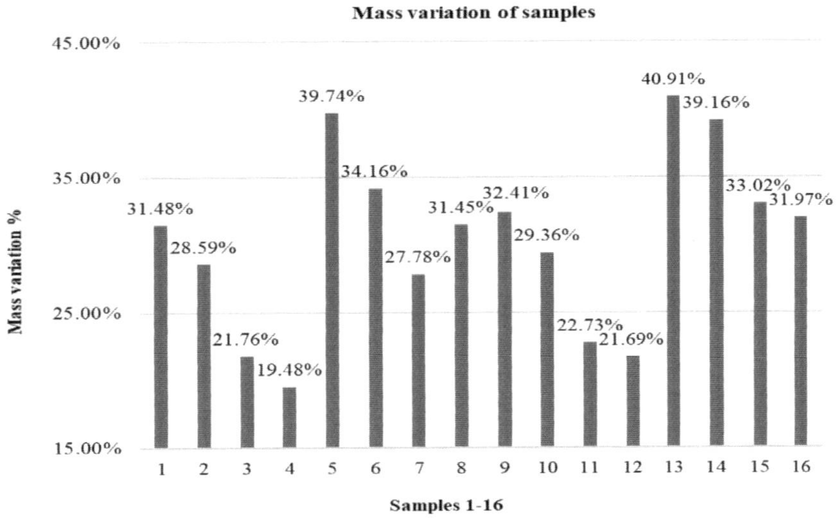

Figure 7. Mass loss after 28 days of curing.

4.2. Experimental values of density after 28 days of curing time

After 28 days the density of the samples was determined according to EN 12390-7 [29], and the values are summarized in table 7.

Table 7. Experimental values for density.

Mixture	Volume [cm^3]	Mass [g]	Density [kg/m^3]
1	3398	1739	512
2	3443	1991	578
3	3443	2330	677

4	3489	2302	660
5	3341	1686	505
6	3477	2091	601
7	3397	2303	678
8	3419	2443	714
9	3364	1960	583
10	3443	2108	612
11	3488	2318	664
12	3546	2369	668
13	3409	1951	572
14	3374	2104	624
15	3394	2422	714
16	3422	2568	750

Analyze the graph from figure 8, it can be seen that the lowest density was obtained for samples 5 and 1, while density with highest value is recorded for sample 16. According to mixture used to cast these two samples, both have the highest amount of water. The differences are in case of sodium silicate and Portland cement to hydrated lime ratio. Sample 16 has 50% much more sodium silicate then the sample 5 and the quantity of Portland cement used is three times much more that it was used in the case of sample 5.

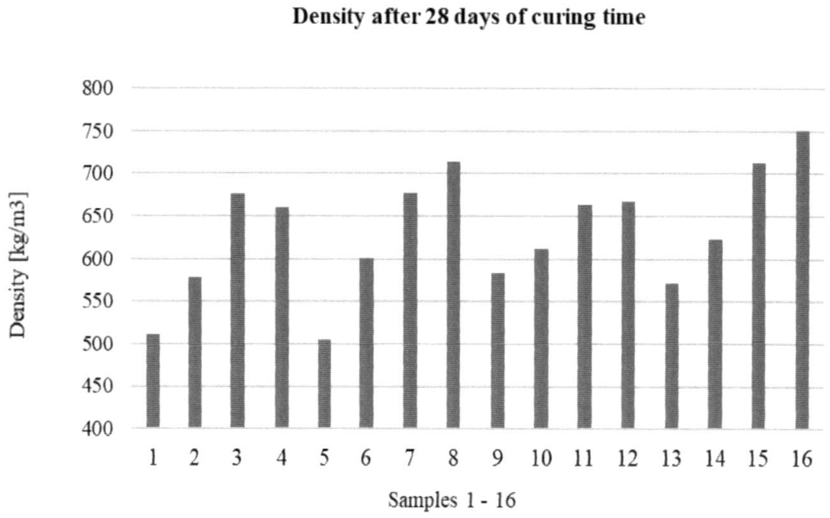

Figure 8. Density of the samples 1 to 16 after 28 days of curing time

4.3. Measurements for determining the heat transfer coefficient

The heat transfer coefficient was determined after 28 days of curing. The measurements were performed using ISOMET 2114 – a laboratory equipment for the materials heat transfer properties measurements, as shown in figure 9.

Figure 9. Heat transfer coefficient measuring.

The experimental values obtained are summarized in table 8, together with the densities values to highlight the interdependence between them. In figure 10, the graph points out, also, the relation between the density and the heat transfer coefficient.

Table 8. Density and heat transfer coefficient.

Mixture	Density [kg/m^3]	Heat transfer coefficient [W/m × K°]
1	512	0.1136
2	578	0.1227
3	677	0.1979
4	660	0.1484
5	505	0.1204
6	601	0.1485
7	678	0.2156
8	714	0.2263
9	583	0.1328
10	612	0.1358
11	664	0.1594
12	668	0.1647
13	572	0.1315
14	624	0.1640
15	714	0.1933
16	750	0.2593

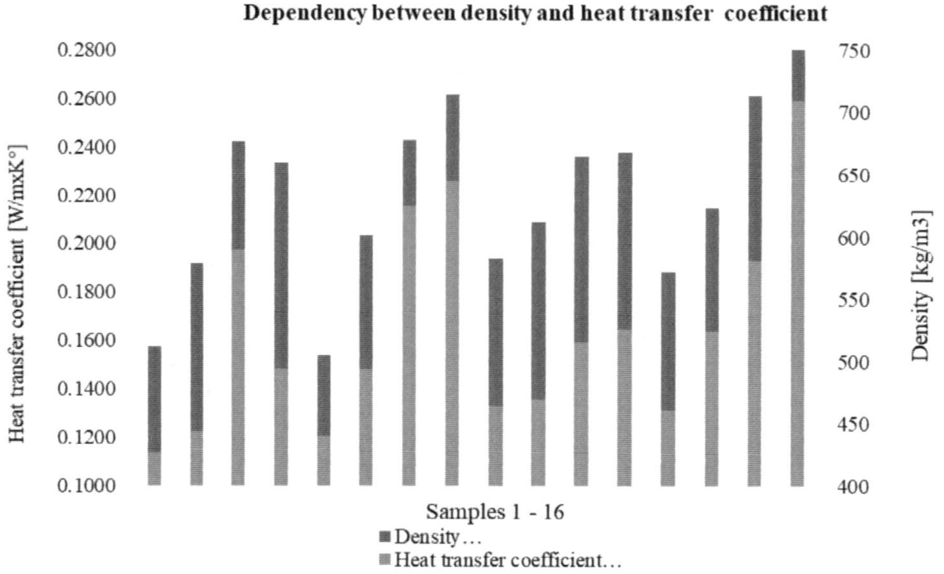

Figure 10. Density and heat transfer coefficient interdependency.

4.4. Determining the compression strength

After 45 days of curing all 16 samples were tested with WAW-600E hydraulic press (figure 11) to obtain the characteristic curve Force-Displacement. The test for determining the compression strength was carried out accordingly to EN 12390-3 [30]. Because hemp concrete is a non-load-bearing material with heat-insulating properties, and its behavior is different from the specimens made of concrete with mineral aggregates, the specimens made of hemp concrete cannot have a point that can be spot on the graph as representing the maximum admissible load. Under the influence of the increasing force, the material deforms plastically without opposing resistance or without having an obvious breaking zone. Therefore. to compare the compressive strength of the samples, the maximum force recorded on each sample at a deformation of 5 mm was used to determine values for compressive strength shown in table 9.

Figure 11. Compression test with WAW-600E hydraulic press.

Table 9. Experimental values for compressive strength.

Sample	Density [kg/m³]	Heat transfer coefficient [W/m × K°]	Compressive strength (45 curing days) [MPa]
1	512	0.1136	0.373
2	578	0.1227	0.433
3	677	0.1979	0.762
4	660	0.1484	0.645
5	505	0.1204	0.308
6	601	0.1485	0.477
7	678	0.2156	1.091
8	714	0.2263	0.792
9	583	0.1328	0.450
10	612	0.1358	0.385
11	664	0.1594	0.711
12	668	0.1647	0.549
13	572	0.1315	0.301
14	624	0.1640	0.349
15	714	0.1933	0.802
16	750	0.259	0.870

In figure 12 is presented the comparison between density, heat transfer coefficient and compressive strength. Analyzing the obtained results, it can be seen that there is a dependency relationship between them. The samples with a lower density, obtain reduced values both for the heat transfer coefficient and for the compression resistance. Any increase in density determines the increase in the others two characteristics.

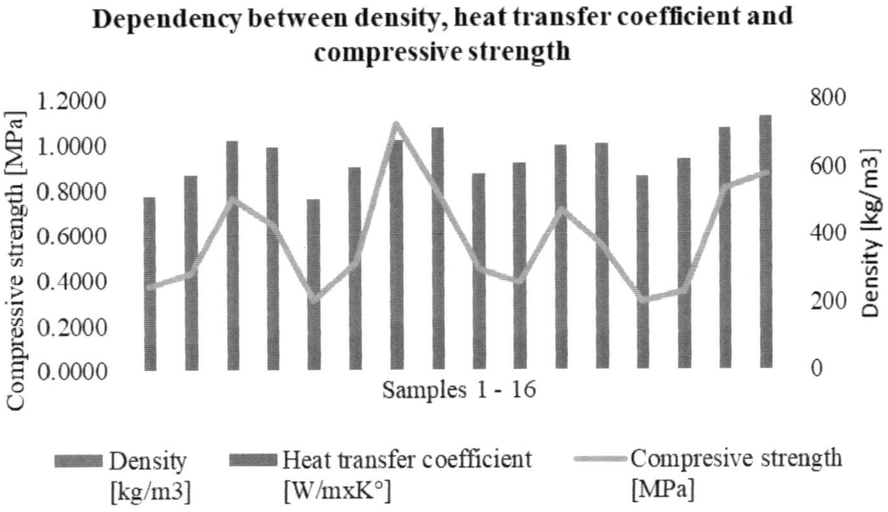

Figure 12. Density, heat transfer coefficient and compressive strength dependency

4.5. Statistical analysis of the obtained results

In table 10 are summarized the independent variables or input factors / parameters and the dependent variables or output response of the system. The statistical analysis will be made for all of the output response to investigate what are the parameters that has a significantly influences on the physico-mechanical properties of the hemp concrete.

Table 10. Experimental values of the system response obtained and related factors/parameters.

Mixture	Independent variables INPUT				Response/Interdependent variables OUTPUT			
	Hydrated lime	Portland Cement	Water	Sodium Silicate	Mass [kg]	Density [kg/m³]	Heat transfer coefficient [W/m x K°]	Compressive Strength [MPa]
1	492	165	657	33	1739	511.836	0.1136	0.373
2	714	165	657	33	1991	578.308	0.1227	0.433
3	492	510	657	33	2330	676.745	0.1979	0.762
4	714	510	657	33	2302	659.844	0.1484	0.645
5	492	165	1224	33	1686	504.579	0.1204	0.308
6	714	165	1224	33	2091	601.342	0.1485	0.477
7	492	510	1224	33	2303	677.881	0.2156	1.091
8	714	510	1224	33	2443	714.454	0.2263	0.792
9	492	165	657	63	1960	582.696	0.1328	0.450
10	714	165	657	63	2108	612.279	0.1358	0.385
11	492	510	657	63	2318	664.473	0.1594	0.711
12	714	510	657	63	2369	668.161	0.1647	0.549
13	492	165	1224	63	1951	572.376	0.1315	0.301
14	714	165	1224	63	2104	623.548	0.1640	0.349
15	492	510	1224	63	2422	713.516	0.1933	0.802
16	714	510	1224	63	2568	750.487	0.1136	0.870

4.5.1. Statistical analysis of the parameters influencing the Mass of the samples after 28 days of curing.
Table 11 shows the values obtained through variance analysis. These values highlight the importance that cement and hydrated lime have for the chosen dependent variables, respectively in the mass variation and implicitly the density, as well as the heat transfer coefficient variation. This is due to the fact that the two mineral binders have the highest volume ratio in the binder mixture to water and sodium silicate.

Table 11. Analysis of variance for Mass.

ANOVA table of mass R-sqrt =.97941; Adj:.93822					
	SS	df	MS	F	p
(1) Hydrated lime	100330.6	1	100330.6	24.9289	0.004130
(2) Cement	733164.1	1	733164.1	182.1678	0.000040

(3) Water	12712.6	1	12712.6	3.1587	0.135672
(4) Silicate	52326.6	1	52326.6	13.0015	0.015449
1 by 2	26325.1	1	26325.1	6.5409	0.050799
1 by 3	11077.6	1	11077.6	2.7524	0.158003
1 by 4	4590.1	1	4590.1	1.1405	0.334375
2 by 3	9168.1	1	9168.1	2.2780	0.191602
2 by 4	6280.6	1	6280.6	1.5605	0.266883
3 by 4	1040.1	1	1040.1	0.2584	0.632834
Error	20123.3	5	4024.7		
Total SS	977138.4	15			

The Pareto diagram (figure 13) shows graphically that the amount of sodium silicate used in the mixture also has an important influence on the mass. The parameter that has a small influence on the mass of the specimens is the ratio of Portland cement to hydrated lime. It can also be seen that the water used is insignificant for the mass of the samples because the amount chemically bound is small, the difference being represented by the water used to obtain the workability of the mixture, water that during the casting process has been removed.

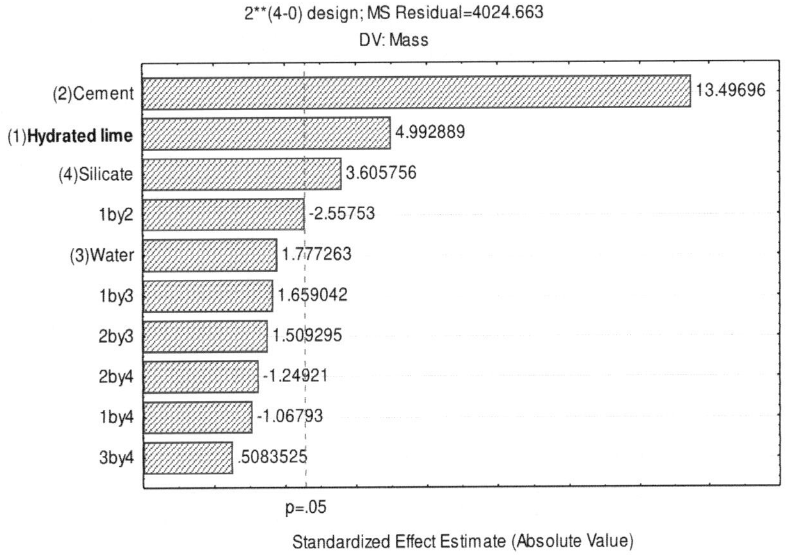

Figure 13. Pareto chart of standardized effects for mass.

4.5.2. Statistical analysis of the parameters influencing the Density of the samples

For hemp concrete mixture investigated, the analysis of variance (table 12) reveals that the major factor that influence the density of the samples is Portland cement. The amount of cement used is mandatory for obtained a desirable value for density.

Table 12. Analysis of variance for Density.

ANOVA table of Density R-sqrt = 0.9826; Adj: 0.94779					
	SS	df	MS	F	p

(1) Hydrated lime	5788.21	1	5788.21	21.9776	0.005395
(2) Cement	55060.51	1	55060.51	209.0628	0.000029
(3) Water	2596.93	1	2596.93	9.8604	0.025662
(4) Silicate	4308.19	1	4308.19	16.3580	0.009879
1 by 2	2108.19	1	2108.19	8.0047	0.036707
1 by 3	1201.25	1	1201.25	4.5611	0.085785
1 by 4	236.34	1	236.34	0.8974	0.386984
2 by 3	1814.52	1	1814.52	6.8897	0.046825
2 by 4	1009.99	1	1009.99	3.8349	0.107523
3 by 4	230.99	1	230.99	0.8771	0.392001
Error	1316.84	5	263.37		
Total SS	75671.97	15			

Although hydrated lime, sodium silicate and water influence the density of the sample, the cement content of each mixture brings in addition to its own weight the influence given by the dependence on the water used for hydration, as shown in figure 14.

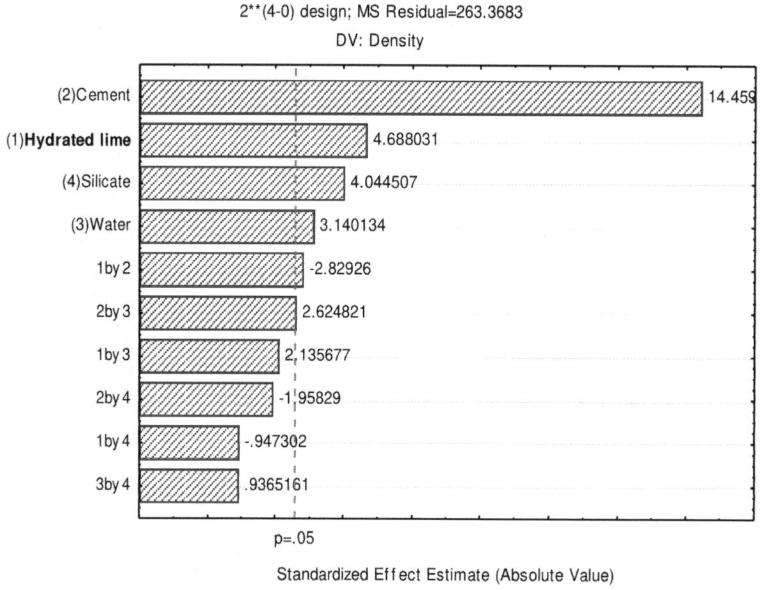

Figure 14. Pareto chart of standardized effects for Density

4.5.3. Statistical analysis of the parameters influencing the Heat transfer coefficient

The analysis of variance (table 13) highlights the major importance that Portland cement has in terms of influencing the heat transfer coefficient of the hemp concrete.

Table 13. Analysis of variance for heat transfer coefficient

ANOVA table of Conductivity heat transfer coefficient R-sqrt = 0.93446; Adj:.80338					
	SS	df	MS	F	p

(1) Hydrated lime	0.000835	1	0.000835	2.13818	0.203527
(2) Cement	0.011257	1	0.011257	28.81901	0.003019
(3) Water	0.007868	1	0.007868	20.14167	0.006471
(4) Silicate	0.000035	1	0.000035	0.09063	0.775492
1 by 2	0.000056	1	0.000056	0.14209	0.721681
1 by 3	0.001580	1	0.001580	4.04504	0.100489
1 by 4	0.000882	1	0.000882	2.25819	0.193225
2 by 3	0.003475	1	0.003475	8.89643	0.030705
2 by 4	0.001253	1	0.001253	3.20815	0.133271
3 by 4	0.000605	1	0.000605	1.54924	0.268407
Error	0.001953	5	0.000391		
Total SS	0.029800	15			

The Pareto's chart (figure 15) revels graphically that a significantly importance has the amount of water. The combined effect of the cement to water ratio is also an important influence parameter. An observation appears here related to the differences between the importance of the factors that influence the heat transfer coefficient and the factors that influence the compressive strength. In the case of the compressive strength, the combined influence of the relationship between the amount of cement and the percentage of water used in the mixture is more important because it is necessary to obtain the total hydration of the cement mass to increase the compression resistance. In the case of the heat transfer coefficient, the importance is rather the separate influence of the cement and the amount of free water which, in addition to the fact that positively influences the workability, at the same time also increases the degree of compaction, which has the effect of increasing the heat transfer coefficient, a characteristic undesirable for hemp concrete.

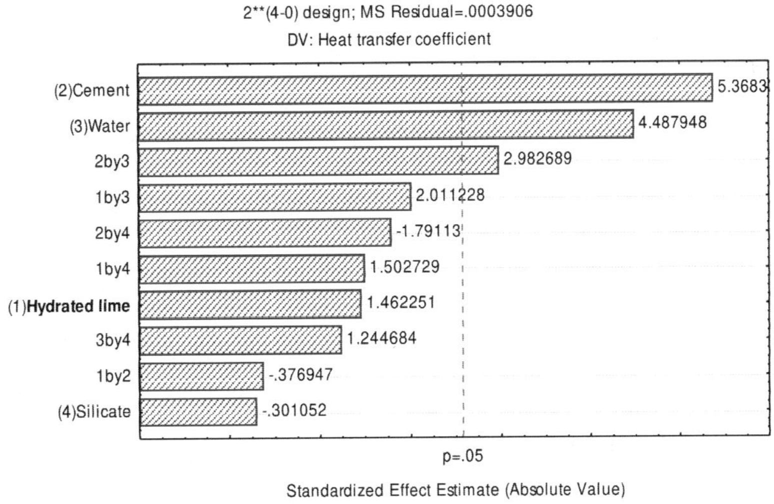

Figure 15. Pareto chart of standardized effects for heat transfer coefficient.

4.5.4. Statistical analysis of the parameters influencing the Compressive strength of samples after 45 days of curing

In the case of compression, the factors that significantly influence its evolution are the cement and its interaction with water (table 14). Being an exothermic process and therefore consuming water during hardening, these interactions of the two parameters are normal.

Table 14. Analysis of variance for Compressive strength.

ANOVA table of Compressive strength R-sqrt = 0.93446; Adj:.80338					
	SS	df	MS	F	p
(1) Hydrated lime	0.000835	1	0.000835	2.13818	0.203527
(2) Cement	0.011257	1	0.011257	28.81901	0.003019
(3) Water	0.007868	1	0.007868	20.14167	0.006471
(4) Silicate	0.000035	1	0.000035	0.09063	0.775492
1 by 2	0.000056	1	0.000056	0.14209	0.721681
1 by 3	0.001580	1	0.001580	4.04504	0.100489
1 by 4	0.000882	1	0.000882	2.25819	0.193225
2 by 3	0.003475	1	0.003475	8.89643	0.030705
2 by 4	0.001253	1	0.001253	3.20815	0.133271
3 by 4	0.000605	1	0.000605	1.54924	0.268407
Error	0.001953	5	0.000391		
Total SS	0.029800	15			

There is certainly a non-linear behavior of this process. which makes water not appear as an important parameter influencing compression. which is not true. Therefore. once the interaction between these two parameters and compression are showed. it can be counted among the factors that influence this mechanical characteristic.

Pareto's chart (figure 16) comes to graphically confirm what was specified above with a probability of 0.95% or with an error of 0.05%. This also shows that, for this series of experiments, the parameters with lower influence are lime and silicate.

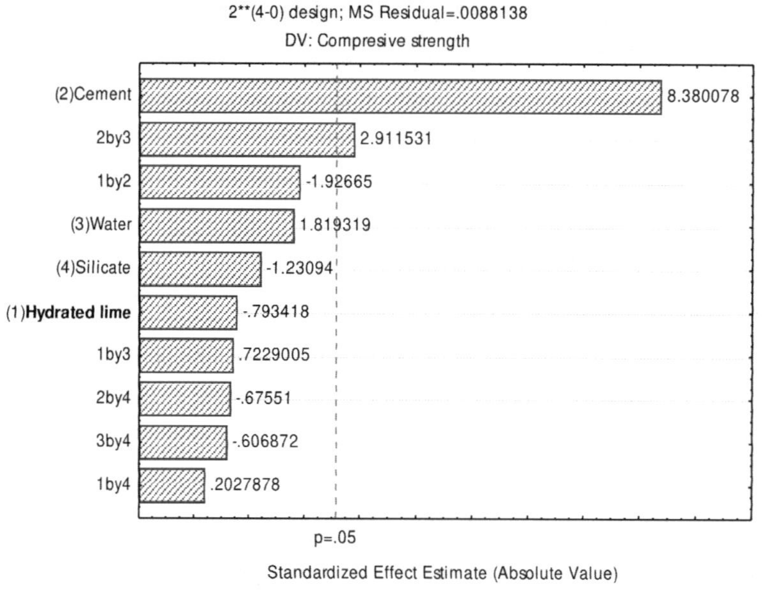

Figure 16. Pareto chart of standardized effects for compressive strength.

5. Conclusions

Advanced design of experiment method helped to rank the analyzed parameters in order to study their influence on the physic-mechanical characteristics of hemp concrete. The heat transfer coefficient is influenced to a large extent by both cement and water and their interaction. This means that the free water from the composition is significant for obtaining a high porosity of the hemp concrete in order to reduce thermal conductivity. The compression strength is significantly influenced only by cement and cement-water interaction. To obtain a better strength on compression stress is mandatory to use water only for hydrate the cement. In the end, the study shows that while mass is influenced by cement, lime and silicate, the density is dependent by all parameters studied.

Also using the method, it can considerably be reduced the number of experiments in order to be able to obtain the same volume of information. The DoE method offers the possibility of obtaining the mathematical relationship of interdependence between the parameters in the mixture so that a material with a certain density or a certain heat transfer coefficient will be designed optimizing the mixture by varying the quantities, depending on the weighting of their influence on the thermo-mechanical characteristics obtained as response to experimental planning.

The aim of the paper was to investigate the influence of parameters define as independent variables on the characteristics of hemp concrete, using a powerful experimental design and analysis tool that allows to understand the behavior of this material, with different percentages and simultaneous configurations of all parameters, which would not have been possible with the classical method using percentage variations of a single factor.

References

[1] Perletti M A 2020 *Sustainable building with hemp. Guide to the use in construction of a natural and innovative material* (Rimini: Maggioli Editor)

[2] Dartois S et al 2017 An iterative micromechanical modeling to estimate the thermal and mechanical properties of polydisperse composites with platy particles: application to anisotropic hemp and lime concretes, *Constr. Build. Mater* **152** 661-671

[3] Page J, Sonebi M, Amziane S, 2017 Design and multi-physicalproperties of a newhybrid hemp-flax composite material, *Constr. Build. Mater* **139** 502–512

[4] Tronet P, Lecompte T, Picandet V, et al, 2016 Study of lime hemp concrete (LHC) mix design casting process and mechanical behaviour, *Cement Concr. Compos.* **67** 60–72

[5] Ruus A, Koosapoeg T, Pau M, Kalamees T and Põldaru M 2021 Influence of production on hemp concrete hygrothermal properties: sorption, water vapour permeability and water absorption – *Journal of Physics: Conf, **2069** 012004*

[6] Sławomir Pochwała, 2020 The Heat Conductivity Properties of Hemp–Lime Composite Material Used in Single-Family Buildings, Materials, building materials, *Energy Proc.* **132** 189–194

[7] Delhomme F, Hajimohammadi A, Almeida A, et al, 2020, Physical properties of Australian hurd used as aggregatefor hemp concrete, *Mater. Today Commun.*, **24** 100986

[8] Fernea R, Manea D L, Plesa L, Iernutan R and Dumitran M, 2019 Acoustic and thermal properties of hemp-cement building materials, *Procedia Manufacturing*, **32** 208-215

[9] Seng B, Magniont C, Lorente S 2019 Characterization of a precast hemp concrete. Part I: physical and thermal properties. *Journal of Building Engineering*, **24** 100540-10.1016

[10] Page J, Sonebi M, Amziane S, 2017 Design and multi-physical properties of a new hybrid hemp-flax composite material, *Constr. Build. Mater.* **139** 502–512

[11] Aytac S 2018 An enviromentally friendly plant in terms of oxigen supply: hemp, *Proocedigs of ICOEST 4th International conference on environmental science and technology*, 19-28 september, Kiev-Ukraine, pp 32-24

[12] Arrigoni R et al 2017 Life cycle assessment of natural building materials: the role of carbonation, mixture components and transport in the environmental impacts of hempcrete blocks, *J. Clean. Prod.* **149** 1051–1061

[13] Faiz Ahmed A T M, Zahidul Islam M et al, 2022 Hemp as a potential raw material toward a sustainable world: A review, *Helyon*, **8(1)** E08753
[14] Ingrao C, Lo Giudice A, Bacenetti J et al, 2015 Energy and environmental assessment of industrial hemp for building applications: a review, *Renew. Sustain. Energy Rev.* **5** 29–42
[15] Magniont C, Escadeillas G, Coutand M and Oms-Multon C, 2012 Use of plant aggregates in building ecomaterials, *European Journal of Environmental and Civil Engineering*, **16:sup1**, s17-s33
[16] Arehart J H, Nelson W S, Srubar W V, 2022 On the theoretical carbon storage and carbon sequestration potential of hempcrete, *J. Clean. Prod.*, **266** 121846
[17] Tarun J, Karade S.R., Singh L.P., 2019 – A review of hemp concrete for building applications, *Journal of Cleaner Production,* **239** 117852
[18] Costantine G, Maalouf C, Moussa T, et al, 2018 Experimental and numerical investigations of thermal performance of a Hemp Lime external building insulation, *Build. Environ*, **131** 140–153
[19] Abdellatef Y., Kavgic M, 2020 Thermal, microstructural and numerical analysis of hempcrete microencapsulated phase change material composites, *Appl. Therm. Eng.* **178** 11552
[20] Sivasubramanian J and Gino Y, 2019 Effect of sodium silicate on hardening property of concrete, MEPCO Schlenk Engineering College, Sivakasi, India, *International Journal of Civil Engineering and Technology*, **10(05)** 642-650.
[21] Hamla W, Benamara D, Noui A, 2022 Statistical modeling of physical and mechanical responses of roller-compacted sand concrete made with ternary sand using the experimental design method, *Constr. Build. Mater.,* **345** 128354
[22] Werdine D, Guilherme A O, Almeida F A, Noronha M L, Gomes G F, 2021 Analysis of the properties of the self-compacting concrete mixed with tire rubber waste based on design of experiments, *Structures,* **33**, 3461-3474
[23] Venkatesan M, Zaib Q, Shah I H, Park H S, 2019 - Optimum utilization of waste foundry sand and fly ash for geopolymer concrete synthesis using D optimal mixture design of experiments, *Resurces, Conservation and Recycling,* **148** 114-123
[24] Panda S and Sarkar P, 2022 - Abrasion resistance of copper slag aggregate concrete designed by Taguchi method, *Materials Today,* **65(2)** 434-441
[25] Bliuc I Lepadatu D et all 2017 Assessment of thermal bridges effect on energy performance and condensation risk in buildings using DoE and RSM methods European Journal of Environmental and Civil Engineering, 21(12) pp 1466-1484
[26] Lepadatu D, Judele L, Antonescu I 2018 Advanced Experimental Design and Optimization Methods for New Building Materials Developed in Civil Engineering. In F. Kongoli, F. Marquis, P. Chen, T. Prikhna, N. Chikhradze (Eds.), *Sustainable Industrial Processing Summit SIPS2018 Volume 6. New and Advanced Materials and Technologies* Montreal, Canada: FLOGEN Star Outreach pp. 263-286
[27] Montgomery D C 2001, *Design and analysis of experiments*, (John Wiley, New York 5th edition)
[28] SR EN 12390-1 Test on hardened concrete. Part 1: Shape, dimensions and other conditions for specimens and patterns
[29] SR EN 12390-7 Tests on hardened concrete. Part 7: Density of hardened concrete
[30] SR EN 12390-3 Testing on hardened concrete. Part 3: Compressive strength of specimens

A non-destructive method for fixing placards with masonry structures

L Bianco[1,2*] **and A Bianco**[3]

[1] Faculty for the Built Environment, University of Malta, Msida MSD 2080, Malta
[2] Faculty of Architecture, University of Architecture, Civil Engineering and Geodesy, Sofia 1164, Bulgaria
[3] Faculty of Engineering, University of Malta, Msida, MSD 2080, Malta

* lino.bianco@um.edu.mt, alessandra.bianco.19@um.edu.mt

Abstract. It is common practice for government agencies to fix signs – and indeed even official temporary notices – onto masonry structures using steel nails. This method of fixing is destructive; it causes irreversible damage to the material fabric, which has a bearing on the overall aesthetic of the building, a scenario which is more acute if the site is of cultural heritage significance. Stones and concrete blocks (locally referred to as concrete bricks), the latter introduced in the 1950s, are the main materials used in masonry construction in Malta. Clay bricks are not utilized, as in Malta there is a blanket prohibition on the extraction of local clay. The main building material used in Malta since time immemorial is Lower Globigerina Limestone. This article puts forward the case for a non-destructive, reversable method to fix notices to building which respects the integrity of the dimension stones. Instead of being hammered into masonry blocks, the proposed removable plugs are installed in the mortar. Their size is relative to the thickness of the mortar bed and the load they are designed to carry, the latter being of negligible importance in the case of lightweight placards. The proposed solution applies equally to other masonry structures, whether erected in dimension or randomly placed stones, concrete blocks or clay bricks, as long as the construction in question uses mortar in the joints between the units.

1. Introduction

A now dated, but still valid, technical report on stone conservation was authored by James Clifton of the National Bureau of Standards, USA [1], a few years prior to the main comprehensive textbook on conservation by Bernard Feilden was published [2]. Citing a publication issued by the Building Research Station with respect to susceptibility of ferrous materials in the original construction of built heritage to corrode and thus induce cracking and subsequent spalling of stonework [3], Clifton called for the use of non-corroding materials such as epoxy-coated steel [4] and non-corroding non-ferrous alloys [3]. In reviewing man-made causes of decay, Feilden discusses damage caused by vibration and atmospheric pollution. In addressing special techniques of repair and structural consolidation, he tackles drilling techniques, notably drilling works and grouting [2]. This paper will address the practice of fixing placards – whether commercial advertisement signs, posters and billboards, or official notices – to masonry structures using steel nails. This is in fact the method used when posting official notices of development applications submitted to the planning regulator in Malta, the main island in the Maltese Archipelago (figure 1a). This practice is destructive and causes irreversible damage to the building fabric, as well as having indisputable aesthetic impacts, especially problematic with culturally significant heritage buildings. Based on a qualitative overview of heritage value typologies covered by leading publications (notably by Mason [5], Yung and Chan [6], de la Torre [7], Doğan [8], Chen and Li [9] and Olukoya [10]), the most important markers of the heritage value of a building is its aesthetic/artistic qualities, followed by its historic, cultural/symbolic, social/community and economic value [11].

The Maltese islands support a building tradition spanning five and a half millennia. Its oldest pre-historic neolithic and megalithic temples were included in the UNESCO's World Heritage List in 1980 "*not only because of their originality, complexity and striking massive proportions, but also because of the considerable technical skill required in their construction*" [12]. With a superficial area of circa 316 km^2, Malta and its dependencies represent the smallest EU Member State and, at the same time, the nation state with the highest density of designated World Heritage sites. Its capital, Valletta, was also listed by UNESCO in its entirety as a World Heritage Site in 1980 [13] (figure 1b). It is one of the most concentrated heritage areas on the globe – 320 cultural monuments within an area of 55 ha inclusive of the St John's Co-Cathedral and the auberges of the Hospitaller Order of St John. This city, founded in 1566, is primarily constructed in Lower Globigerina Limestone (LGL), described by the English poet Samuel Taylor Coleridge (1772–1834) in 1804 as a "*sand-free stone*" [14]. It is "*remarkable for its softness; it is worked easily, but it is not strong enough against moistures* [sic] *and the sea-breeze*", an observation included in a publication issued in Lyon in 1536 authored by Johannes Quintinus D'Autun (1500–1561) [15]. Figure 2 illustrates the typical practice of using steel nails on cultural heritage buildings, in this case to post a notice of a planning application in Valletta. The main objective of this paper is to put forward a prototype plug to fix signs onto masonry which does not cause irreversible damage to the material fabric and thus it respects the integrity of the dimension stones notably of cultural heritage significance.

Figure 1. The Maltese Archipelago: (a) the main habitable islands; (b) Valletta, including the location of St John's Co-Cathedral and the auberges of the Hospitaller Order of St John (© Google Earth). Location of case study shown in figure 2 is circled in green.

2. Globigerina Limestone Formation

2.1. Lithostratigraphy

Lower Globigerina Limestone (LGL), the oldest member of the Globigerina Limestone Formation, is Aquitanian in age (23 million to 20.4 million years ago) [16, 17]. A comprehensive description of its lithostratigraphy is given in Baldassini *et al.* [18]. It is composed of massively bedded, biodetrital limestone consisting of globigerinid planktonic foraminifera. In contrast with the upper part of this member, the lower part is coarser-grained and strongly bioturbated. Pectinid bivalves and echinoids such as Schizaster are present in the former, and bryozoans and the echinoid Scutella in the latter [18, 19].

(a) (c)

(b) (d)

Figure 2. Development planning application at 99 Archbishop's Street on the corner with 81 Bakery Street, Valletta. Streetscape along (a) Archbishop's Street and (b) Bakery Street with site notice indicated by the authors in red and yellow markers respectively [20]; site notice (c) as on 12 January 2022 [20], and (d) detail marked in (c) as on 12 October 2022 – note the rusting steel nail.

In the building trade, LGL is graded in terms of first- and second-quality lithotypes. Good or first-quality LGL has a characteristic resonant vibration – it 'rings' on being struck with a hammer [21]. The inferior second-quality LGL, known as '*sol*' (also written as '*soll*') occurs in two varieties: sol aħmar (red sol) and sol ikħal (blue sol); the former is used in the building industry, often at a height of at least 1.2 m above the damp-proof course. In this paper, inferior LGL refers to sol aħmar.

2.2. *Petrophysical characteristics*

Microphotographs illustrating the texture, cement fabric and pores of both LGL lithotypes, studied using a Hitachi S-520 scanning electron microscope (SEM), are reproduced in figure 3. The first-quality lithotype has well defined physico-mechanical interlocking; its pore structure and fine-grained sparry calcite, which fills the inter-particle voids, cements most allochems (figure 3a). This interlocking physico-mechanical bond, pore structure and fine-grained cement account for its better durability compared to the second-quality lithotype (figure 3b).

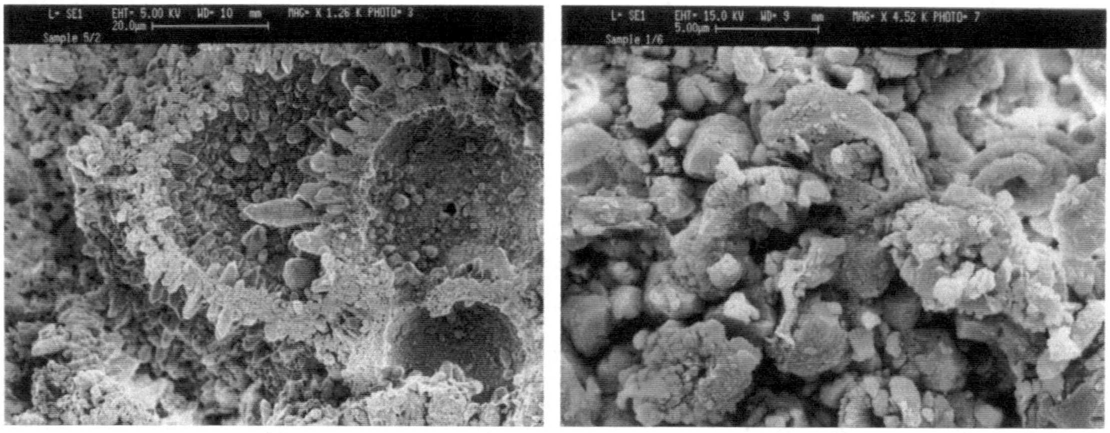

(a) (b)

Figure 3. SEM images illustrating the pore structure, physico-mechanical interlocking and fine-grained sparry calcite cement in (a) the first-quality lithotype and (b) the second-quality lithotype [22].

With respect to LGL, the uniaxial compressive strength in its dry and saturated state is in the range of 15 N/mm^2 to 32.9 N/mm^2 and 9.1 N/mm^2 to 16.3 N/mm^2, respectively [23]. For inferior LGL, the mean dry compressive strength (f_k) is characteristically higher and porosity is lower [24]. In the terms set out by the Centre Technique de Matériaux Naturels de Construction [25, 26], LGL is unquestionably not a hard stone ($f_k > 40$ N/mm^2); some lithostratigraphic beds may be considered compact (10 N/mm^2 $< f_k \leq 40$ N/mm^2) rather than soft stone ($f_k \leq 10$ N/mm^2).

3. Proposed design solution

3.1. *The issue*

Where nails have been used to fix something to the surface of a building, there are two options when the item is due to be removed: (i) remove the nail or (ii) drive it further in. The former case, which is the most common, involves driving a cat's claw or hammer claw under the nail's head and levering it out. The latter case is more damaging in the long term, as additional stresses are generated due to corrosion of the nail.

Although nails have a narrow end, the pullout force (F_p) required to pluck them out of LGL generates a combination of tensile and shear stresses, resulting in a cone of the fabric being removed. F_p is directly proportional to the compressive strength of the limestone (σ_m) thus:

$$F_p = \sigma_m \, a_c \tag{1}$$

where a_c is the surface area of the conic frustum.

The greater the pullout resistance, the greater the pullout force, resulting in a conic frustum with a larger surface area, resulting in greater irreversible damage to the fabric. Furthermore, given LGL's physico-mechanical characteristics of interlocking and the fine-grained sparry calcite cement (figure 3), the surface area of the conic frustum generated in the first-quality lithotype is greater than in second-grade limestone. Therefore, more damage is caused to cultural heritage buildings predominantly constructed from the high-grade lithotype.

3.2. Design proposal

Historical masonry buildings erected in LGL dimension stones make use of lime-based mortar to level and fill in the spaces between the blocks [27]. The thickness of a mortar joint in such building construction is approximately 5 mm. The performance of the mortar is based on the relative bulk density and the 'free lime' content [28]. Lime-based mortar is not used to cement the joint but to provide a cushion between the blocks.

The design proposed in this study is an anchor plug which is placed in a hole drilled in the mortar of bed and/or head joints and can be removed when required. The concept is based on Rawlbolt®, the popular expanding shield anchor bolt widely used in the building industry. Named after John Joseph Rawlings (1860–1942), this proprietary general-purpose bolt is typically used to a recommended torque, depending on masonry type, where significant load-carrying capacity is required. This is not the case in this study, where the fixings are lightweight. When installed in masonry, Rawlbolts® generate a combination of tensile and shear stresses.

The proposed design – the RR-Plug – is a removable and reusable stainless steel plug with an expanding sleeve, a ferrule with a built-in flange and an angle-tapered nut, the latter ensuring maximum expansion on tightening the bolt (figure 4). The drill bit is slightly larger than the shield of the plug and can therefore be loosely placed in the hole ($\phi_h > \phi_p$, where ϕ_h is the diameter of the blind hole in the mortar and ϕ_p the diameter of the plug) once it has been cleared of debris. Once fully assembled, when the bolt and washer are tightened, the tapered nut is pulled towards the plug's shield via the bolt's thread; the angle-tapered nut does not lock in the shield on tightening. This expands the shield, securing the plug in its orifice. For extraction, the bolt is untightened and gently hammered in so that the nut angle moves away from the shield. The purpose of the inbuilt flange is to minimize and absorb most of the stresses transmitted to the masonry in which the plug is set (when it needs to be gently hammered to extract it). On hammering, the flange helps minimise the damage to the surrounding fabric (ϕ_f, the diameter of the flange, must be at least $3\phi_p$).

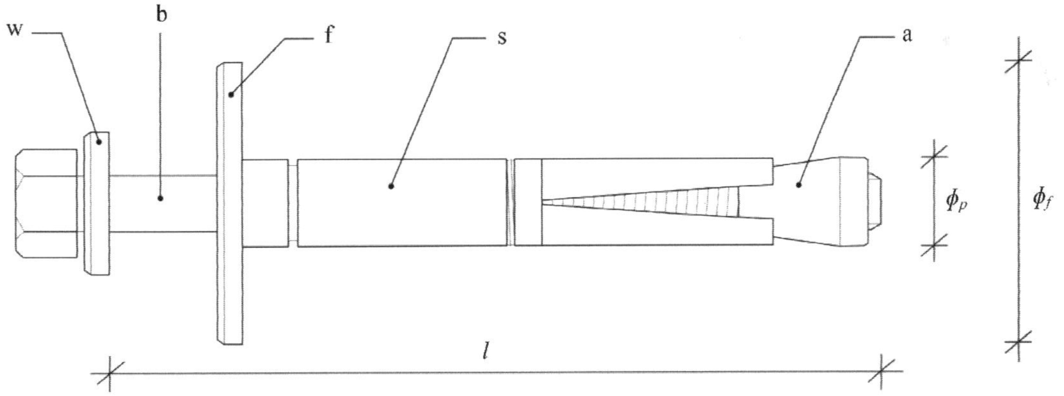

b : bolt
w : washer
f : ferrule with in-built flange
s : shield
a : angle-tapered nut

ϕ_p : overall diameter of RR-Plug
ϕ_f : diameter of flange
l : length of RR-Plug

Figure 4. Conceptual sketch of proposed design of RR-Plug (not to scale). The RR-Plug's main components include a bolt and washer, a shield (which includes a ferrule with a built-in flange) and an angle-tapered nut.

On hammering, the stress, σ, generated on the fabric is given by:

$$\sigma = F_h a^{-1} \quad (2)$$

where F_h is the hammering force and a is the area on which the F_h is transmitted.

The surface area of (i) the flange (a_f), (ii) the mortar joint covered by the flange (a_m) and (iii) the drilled hole (a_h) are related as follows:

$$a_f = 0.25\ \Pi\ \phi_f^2 \quad (3)$$

$$a_m \approx \phi_h \phi_f \quad (4)$$

$$a_h = 0.25\ \Pi\ \phi_h^2 \quad (5)$$

If the mortar is level with the masonry,

$$a = 0.25\ \Pi\ (\phi_f^2 - \phi_h^2) \quad (6)$$

and hence,

$$\sigma = F_h\ [0.25\ \Pi\ (\phi_f^2 - \phi_h^2)]^{-1} \quad (7)$$

If the mortar is recessed in the masonry,

$$a \approx \phi_f\ [0.25\ \Pi\ (\phi_f - \phi_h)] \quad (8)$$

and hence,

$$\sigma \approx F_h\ [\phi_f\ (0.25\ \Pi\ \phi_f - \phi_h)]^{-1} \quad (9)$$

Thus, in the case of Malta, if $\phi_h = 0.005$ m and $\phi_f = 0.015$ m, then the stress, σ, is over 50 % greater in recessed mortar than when the mortar is level with the masonry. Thus, it is recommended that prior to installation the mortar joint is rendered, with a reversible mix compatible with the existing fabric, to a relatively flat surface with the dimension stones in order to generate a uniformly distributed stress on the fabric when hammering.

4. Results and discussion

The installation and the eventual extraction of the RR-Plug each involve three stages, as set out below and illustrated in figure 5.

To install:

1. Drill a blind hole with a diameter equivalent to the thickness of the mortar and to a depth of at least 20 % longer than the length of the plug. Thoroughly clean the hole from debris (figure 5a).
2. Insert the RR-Plug in the cleaned blind hole until the flange of the ferrule is level with the surface (figure 5b).
3. Tighten the bolt (figure 5c).

To extract:

1. Loosen the bolt (figure 5d).
2. Gently hammer the bolt in (figure 5e).
3. Remove the RR-Plug (figure 5f).

After extraction, the hole must be thoroughly cleaned of any debris and infilled with a compatible mortar. In the case of LGL, lime-based mortar is compatible.

The hole can be drilled using a portable electrical drill. A drill of up to 0.5 hp is safe both in terms of damage to the fabric and health and safety; live electricity is a hazard in drilling work. If skillfully used, this kind of tool will cause less damage, but attention should be paid to the power of the drill. The principle is simple: the faster the rate of penetration, the more accurate the hole is likely to be [2]. High-speed rotary–percussive drills have a fast rate of penetration.

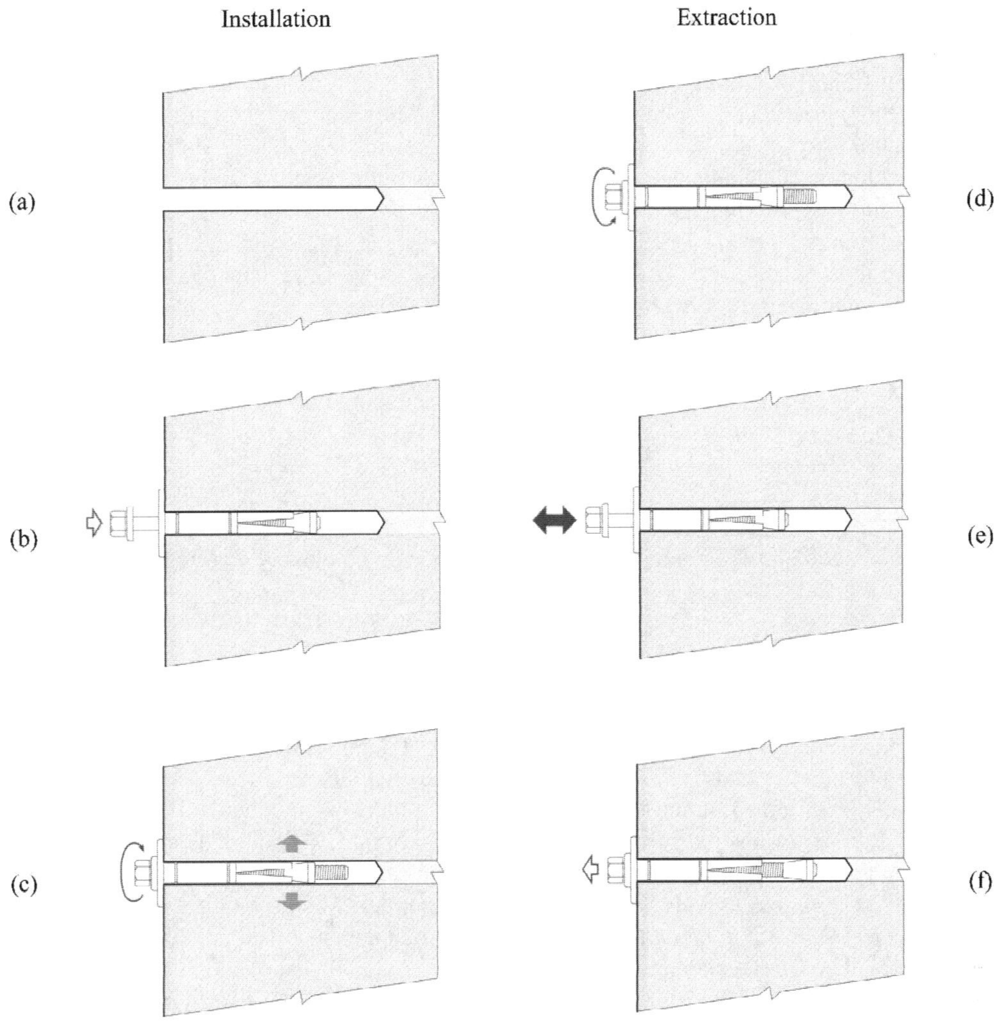

Figure 5. Installation (a to c) and extraction (d to f) guides of a RR-Plug (conceptual; not to scale).

It could be argued that this method is not fully non-destructive, as it does tamper with the mortar, permanently altering it in the affected area. However, in contrast with the current practice, this method does not affect the fabric of the dimension stones, and the mortar can be easily replaced with a compatible mix akin to the original – a common intervention in building construction.

5. Conclusions

Using nails to fix placards to the masonry of heritage buildings is a destructive practice which causes irreversible damage to the built fabric. An alternative method is to use a removable, reusable plug – the RR-Plug – which can be installed in a hole drilled in the mortar of the joints between the blocks. The size of the plug depends on (i) the thickness of the mortar joint (the dimeter of the shield must be slightly less than the thickness of the mortar) and (ii) the load/weight of the placard, which is usually negligible in the case of planning application notices affixed to buildings where works are proposed.

The installation of the RR-Plug is completely reversible without causing any damage to the masonry blocks, and it does not affect the geological fabric of the dimension stones. Although this method does impinge on the integrity of the mortar, a manufactured mix, it does not compromise the integrity of the matrix of the geological fabric. On extraction, the holes drilled for the plugs are infilled with a mix compatible with original mortar. In the case of LGL dimension stone, a lime-based mix should be used. This method can be used in any building constructed from masonry blocks with mortar joints, regardless of the stone used.

Philosophies of conservation, along with practices aimed at preserving the built environment have a long history within Europe, but they do change with the prevailing ideology of the times [29]. While conservation traditionally aimed to retain the memory of the past intact, current philosophy is grounded more in preserving values. The proposed method of using removable and reusable plugs to fix notices to buildings with mortar joints ensures that the heritage values relating to the building fabric of cultural monuments are respected. It eliminates the irreversible damage caused by the current practice and has no impact on the overall aesthetics of the monument.

The RR-Plug is not feasible to use in masonry construction with exceedingly thin mortar joints. Further research is required to develop a non-destructive method to fix signs in such construction whereby the installation and extraction technique does not impinge on the material fabric of the host dimension stones.

Acknowledgements
The authors would also like to thank Professor Inġ. Joseph Buhagiar (Head of the Department of Metallurgy and Materials Engineering, Faculty of Engineering, University of Malta), for his feedback on an earlier draft of this article.

References

[1] Clifton J R 1980 *Stone Consolidating Materials – A status report*. National Bureau of Standards Technical Note 118 (Washington: US Government Printing Office)
[2] Feilden B M 1982 *Conservation of Historic Buildings* (London: Butterworth Scientific)
[3] Building Research Station 1975 *Decay and Conservation of Stone Masonry*. Digest 177 (Garston, England: Building Research Station)
[4] Clifton J R 1976 Protection of reinforcing bars with organic coatings. *Materials Performance* **15**(5) 14–17
[5] Mason R 2002 Assessing values in conservation planning: Methodological issues and choices. *Assessing the values of cultural heritage: Research report*, de la Torre, M. (ed.) (Los Angeles: The Getty Conservation Institute) 5–13
[6] Yung E H K, Chan E H W 2013 Evaluation for the conservation of historic buildings. *Facilities* **31**(11-12) 542–564
[7] de la Torre M 2013 Values and heritage conservation. *Heritage & Society* **6**(2) 155–166
[8] Doğan H A 2020 Perception of the modern movement in architecture as cultural heritage. *Art History & Criticism / Meno istorija ir kritika* **16**(1) 65–78
[9] Chen D, Li J 2021 Process-led value elicitation within built heritage management: A systematic literature review between 2010 and 2020. *Journal of Architectural Conservation* **27**(1-2) 1-16
[10] Olukoya O A P 2021 Framing the values of vernacular architecture for a value-based conservation: A conceptual framework. *Sustainability* **13**(9) 4974
[11] Zammit N, Bianco L 2022 Computation of Heritage Values: Towards a holistic method to assess built heritage. *Heritage and Sustainable Development* **4**(2) 101–110
[12] UNESCO 1980 *Megalithic Temples of Malta* (https://whc.unesco.org/en/list/132/, retrieved 14 August 2022)
[13] UNESCO 1980 *City of Valletta* (https://whc.unesco.org/en/list/131/, retrieved 14 August 2022).
[14] Coleridge S T 1962 *The Notebooks of Samuel Taylor Coleridge*. 2nd ed.; Coburn, K., Ed.; Volume 1804–1808 (London:Routledge & Kegan Paul)
[15] Quintin d'Autin J 1980 *The Earliest Description of Malta* (Tas-Sliema, Malta: DeBono Enterprises)
[16] Giannelli L, Salvatorini G 1972 I foraminiferi planctonici dei sediment terziari dell'arcipelago

Maltese. I. Biostratigrafia dell Globigerina Limestone. *Atti Della Soc. Toscana Sci. Nat. Mem. Ser. A* **79** 49–74
[17] Felix R 1973 *Oligo-Miocene Stratigraphy of Malta and Gozo* (Wageningen, The Netherlands: H. Veenman and B.V. Zonen)
[18] Baldassini N, Mazzei R, Foresi L M, Riforgiato F, Salvatorini G 2013 Calcareous plankton bio-chronostratigraphy of the Maltese Lower Globigerina Limestone member. *Acta Geol. Pol.* **63** 105–135
[19] Foresi L M, Mazzei R, Salvatorini G, Donia F 2007 Biostratigraphy and chronostratigraphy of the Maltese Lower Globigerina Limestone Member (Globigerina Limestone Formation): New preliminary data based on calcareous plankton. *Boll. Della. Soc. Paleontol. Ital.* **46** 175–181
[20] Planning Authority 2022 *PA Case Details: PA/09046/21* (https://www.pa.org.mt/en/pacasedetails?CaseType=PA/09046/21, retrieved 12 October 2022)
[21] Bondin S 1992 Former President of the Softstone Quarry Association, Mqabba, Malta. Personal Communication
[22] Bianco L 2021 Geochemistry, mineralogy and textural properties of the Lower Globigerina Limestone Used in the built heritage. *Minerals* **11**(7): 740. https://doi.org/10.3390/min11070740
[23] Cachia J 1985 *The mechanical and physical properties of Globigerina Limestone as used in local masonry construction* (University of Malta, B E & A (Hons) dissertation)
[24] Bianco L 1993 *Some factors controlling the quality of Lower Globigerina building stone of Malta* (University of Leicester, MSc dissertation)
[25] Centre Technique de Matériaux Naturels de Construction 2015 *Study on characteristic compressive strength of natural stone masonry made with general purpose mortar* (Paris)
[26] Camilleri D H 2019 Malta's Heritage in Stone: From temple builders to Eurocodes 6/8. *MInt*, **31**(2): 49–67
[27] Bianco L, Cardona K 2020 Seventeenth-century building engineering in the central Mediterranean: a case study from Malta. *Terra Sebus. Acta Musei Sabesiensis* **12** 329–353
[28] Hughes J Q 1967 *The Building of Malta during the Period of the Knights of St John of Jerusalem, 1530–1795* (London, Alec Tiranti)
[29] Jokilehto J 1999 *A History of Architectural Conservation* (Oxford: Butterworth-Heinemann).

Perlite concrete: a review

A Cojocaru[1*], D N Isopescu[1*] and S G Maxineasa[1]

[1] "Gheorghe Asachi" Technical University of Iasi, Faculty of Civil Engineering and Building Services, Department of Civil and Industrial Engineering, Iasi, Romania

* alexandra.cojocaru@student.tuiasi.ro

Abstract. The thermal resistance of concrete significantly influences the energy consumption of the building. Consequently, concrete with high thermal resistance can contribute to the envelope's energy efficiency. In order to obtain the thermal conductivity performance required for the passive house standard using the principles of sustainable development, it is essential to analyze the thermal improvement solutions of the concrete with a low impact on the environment. For this reason, it is necessary to analyze methods of improving concrete's thermal performance by replacing aggregates with ecological materials like perlite. Perlite is a common term used for a glassy natural volcanic rock used in construction in various mixtures for its fire resistance, low density, and thermal and acoustic improvement properties. While processing raw perlite (cutting, grinding, and fractionation to various sizes), large amounts of powder are collected and stored as waste. Currently, only limited amounts of waste perlite powder are recycled in the construction industry, which urges the efficient use of this waste to limit the impact on the environment. There are various studies about concrete mixtures with perlite from the point of view of structural strength. However, the results are contradictory, influenced by the size of used particles and the replacement levels in the mixture. This article aims to present the current state of knowledge regarding the characteristics of the mixture between concrete and perlite.

1. Introduction

The built environment plays a critical role in shaping our society and economy, but it also significantly impacts the environment. Buildings and infrastructure consume vast amounts of energy and natural resources and are responsible for a significant portion of global carbon emissions. According to various studies, the built environment is responsible for more than 40% of total energy consumption, 40% of energy-related CO_2, and more than 8% of the total CO_2 generation annually [1-3]. This is a significant concern as these emissions contribute to global warming and climate change. Without effective change, this negative impact can increase up to 25% of the total CO_2 production by 2050 [4]. Those values have a seriously negative impact on the environment and urge new solutions for the building materials as elements that significantly influence the overall thermal transmittance of the building envelope [5]. The thermal transmittance, also known as the U-value of a building's envelope, is a measure of the heat loss through the building envelope, and it is affected by the materials used in construction. Therefore, using sustainable building materials with lower U-value can significantly reduce energy consumption and carbon emissions. Sustainable materials have economic and environmental benefits in the construction industry as the key to reducing emissions related to large quantities of energy inputs and using natural resources [6-8].

Additional benefits of proper construction materials are less spoilage and waste materials [5]. Future modern building materials should have improved functional performance and must reflect the current need to lower the ecological footprint of the materials by preserving natural resources [9]. To preserve natural resources, it can be a beneficial approach to analyze solutions to substitute natural aggregates in the concrete mixture. [10-12]. Overall, sustainable building materials have the potential to mitigate the negative impact of the built environment on the environment and must be given more consideration in the construction industry.

Among thermal use, perlite is one of the outstanding materials that can have numerous applications in new building materials. There are various articles regarding expanded perlite as a part of building material. A review made in 1991 [13] and a review made in 2016 [14] list construction applications: Portland cement-perlite-based plaster, Gypsum perlite-based plaster, perlite concrete, lightweight tile mortar, lightweight composites based on cement/gypsum plaster and perlite binder, and insulation applications also. Even if the articles are very detailed regarding the characteristics of perlite, the literature needs a state-of-the-art update that can summarize the previous studies on perlite concrete. Using perlite as an aggregate in concrete can significantly reduce the weight of the concrete and improve its thermal insulation properties. Additionally, perlite concrete has been found to have good workability and high compressive strength. However, there are also challenges associated with the use of perlite concrete, such as its relatively high water absorption and low durability.

The review aims to comprehensively understand the properties and behavior of perlite concrete and identify potential areas for further research. Therefore, this paper presents the previous findings correlated to the mixture between concrete and perlite to behave as a guide support for upcoming studies.

2. Perlite

The perlite used in construction is a naturally occurring alumino-siliceous amorphous volcanic material obtained from crude perlite rock, found in several countries, such as Italy, Hungary, Turkey, Greece, Japan, and the USA [14-17]. Global perlite production is growing, simulated by increasing demand and improved capacity introductions. This results in relatively large quantities (5-10%) of waste perlite with under 200 lm particles, hard to dispose of in dry conditions. The limited storage capacity in production zones for this waste makes it necessary to use it as an aggregate in construction.

When raw perlite rock is subjected to temperatures of 870 to 1200 ˚C, it expands to 15-35 times its original volume. It transforms into a soft bulk-density cellular material with countless bubbles due to the 2-5% of water in its composition. Perlite is characterized by thick and thin laminar sheets when examined with a microscope. The color of crude perlite rock is light grey to glossy black, and after the expansion, it transforms from snowy white to greyish white [18-21].

Perlite is a unique and versatile mineral widely used in various industrial applications due to its distinct properties. One of the most notable characteristics of perlite is its foamy-like microstructure, which gives it a low density of around 240 kg/m3. This low density, with its good thermal insulation properties and high porosity, makes it an ideal choice as a lightweight mineral filler in many applications [22-36]. From a chemical point of view, perlite is composed primarily of silicon dioxide (SiO2) at an average of approximately 75%, with smaller amounts of aluminum oxide (14.8%), and various metal oxides such as potassium (4.8%), sodium (2.9%), calcium (0.9%), magnesium (0.1%), and R2O3 (1.5%), as well as water (3.0%) [13].

In addition to the chemical and microstructural properties, perlite also possesses various physical characteristics that contribute to its versatility as a mineral filler. These include a density range of 32-400 kg/m3, which is relatively low compared to many other minerals [14]. Additionally, perlite has a pH range of 6.5-8 and a specific gravity of 2.2-2.4 [14]. The mineral has a water retention capacity of 35-50%, which is relatively high compared to other minerals [14]. The thermal conductivity of perlite is also relatively low, with a value of 0.04-0.06 W/mK at 24°C [14]. Additionally, perlite has a specific heat of 387 J/kgK, a softening point of 871-1093°C, and a melting point of 1260-1343°C [14].

These characteristics make perlite an ideal material for the construction industry, as it can be used as an aggregate and a binder in lightweight composites. Perlite is a natural product, ecological in the application, non-polluting, and free of organic substances. The high thermal insulation properties and low density of perlite make it a good choice for thermal insulation applications. Its high porosity and chemical composition make it suitable for construction applications such as insulation and lightweight concrete. Overall, perlite is a versatile and sustainable material that can play an essential role in reducing the environmental impact of the built environment.

3. Perlite concrete

Numerous studies in the literature have investigated the use of expanded perlite in building materials technology, specifically the effects of expanded perlite on the properties of concrete. These studies have shown that replacing expanded perlite in concrete can significantly improve the properties of the concrete, such as its weight and thermal insulation [26-28, 37-45]. During earthquakes, high buildings have been affected in most cases because of the higher unit weight of concrete, which led to the desire to achieve the lightweight property of concrete. Replacing expanded perlite in concrete may respond to this need due to its lightweight properties [46].

Furthermore, because it is widely available (approximately 2.53 and 2.68 million tons of perlite are produced yearly around the world) and low-cost material, its use in concrete is conducive to cost control [14, 47-48]. Its lightweight properties and low cost make it an attractive alternative to traditional concrete in high-rise buildings and other structures.

The use of expanded perlite as a replacement for fine aggregate in concrete has been the subject of numerous studies in the literature. These studies have shown that the various replacement ratios of fine aggregate with expanded perlite give various contradictory properties. The roles of lightweight's aggregates, either natural or artificial, are multiple: they provide volume stability and resistance to wear [13]. Perlite can be used in construction as an aggregate with cement, bitumen, lime, gypsum, or other binders to form concrete or mortar. Furthermore, a complete analysis of the properties will help better understand the effect of perlite in the concrete mixture. As mentioned before, the studies give contradictory results influenced by the size of used particles and the replacement levels in the mixture.

To further investigate the effects of perlite on the properties of concrete, it is essential to conduct studies that consider a range of replacement ratios, particle sizes, and curing methods. Additionally, it is crucial to evaluate the properties of perlite concrete, such as compressive strength, modulus of elasticity, thermal conductivity, and water absorption, to understand the behavior of perlite concrete fully.

The use of perlite in concrete mixtures can significantly impact the final product's workability. Workability is an essential aspect of concrete as it determines the ease with which the mixture can be placed, compacted, and finished. The higher workability of concrete leads to better compaction and less segregation, resulting in a more homogeneous and durable final product. From the point of view of workability, most studies show that using perlite helps to increase it, related to the increase in air content. A study by Karakoç and Demirboga [49] investigated the effect of replacing fine natural aggregate with expanded perlite of 2mm on the workability of concrete. The study found that using perlite as a partial substitute for natural aggregate increased the workability of the concrete. Specifically, the study reported that by replacing fine natural aggregate with expanded perlite of 2mm in 7.5%, 15%, 22.5%, and 30% volume, the workability of the concrete was improved significantly [49].

Studies reported increments in the slump height as higher workability of concrete mixture by replacing up to 50% of the volume of natural aggregate with expanded perlite (size 0.15– 11 mm [50] and 77.8 lm [51]). Contradictory, some studies report a workability reduction for the mixture with a maximum 15% volume substitute ratio by expanded perlite of 0-4mm [52, 53]. As the setting time of perlite concrete is reportedly on the rise, chemical additives are no longer necessary and can be replaced by expanded perlite as retarding material. This can be beneficial when the concrete needs to be placed in difficult or inaccessible areas or the curing process needs to be slowed down to improve the final product.

The incorporation of perlite in concrete mixtures can significantly impact the density of the final product. Due to the perlite's light density, the fresh density measurements of the perlite concrete studies account for a reduction of up to 65.46% if the whole volume of natural sand were replaced by expanded perlite [26]. However, the effect of perlite on density can vary depending on the substitution ratio and the size of the perlite used.

Türkmen and Kantarcı [53] reported a 0.74%, 0.91%, and 0.95% reduction in the fresh unit weight by partially substituting natural sand with expanded perlite (size 0–4 mm), by volume, at 5%, 10%, and 15%, respectively, in contradictory with other studies that show a linear reduction of the fresh density of concrete mixtures including from 5% to 100% perlite [50, 52]. Studies also show a reduction in the 28-day unit weight of concretes by partially substituting natural sand, by volume, with perlite. The reduction linearly increases while substituting more than 15% natural sand with perlite. Under 15% substitution of volume, the reduction is less than 1.36%, as Türkmen and Kantarcı [52, 53] reported in their studies by analyzing the mixture of concrete with perlite (size 0-4mm).

From the density point of view, replacing over 50% of natural sand with perlite decreases the density up to 500kg/m3 due to the low specific gravity, which reduces the mixture's unit weight [25]. The low density is the most valuable part of perlite addition, although it leads to low mechanical properties [22-36]. Thus, it is essential to conduct thorough laboratory testing to determine the optimal perlite ratio for specific applications and to balance the density and mechanical properties.

Compressive strength is an essential property of concrete as a building material, as it determines the ability of the concrete to withstand compressive loads. Several studies have investigated the impact of perlite on the compressive strength of concrete. The results of these studies have shown a significant reduction in the 28-day compressive strength by partially substituting natural sand with perlite [49, 51]. For example, partially substituting natural aggregate with perlite (size 0.15–11 mm) at levels of 10%, 20%, 30%, 40%, and 50% affects the compressive strength of concrete with 39.8%, 63.33%, 80.69%, 84.29%, and 90.58% [50].

However, when smaller amounts of perlite were used, such as 5%, 10%, and 15%, the reduction in 28-day compressive strength was relatively minor, with reductions of only 1%, 3%, and 7% observed at the respective perlite content levels [53]. Expanded perlite used as a part of the aggregate in the concrete mixture leads to an increase in the porosity that leads to a lowering of the density and reduces the strength [22-36, 54]. Moreover, water absorption increase with higher levels of perlite in the mixture, which can further weaken the concrete.

The use of expanded perlite as a replacement for natural aggregate in concrete mixtures can have a significant impact on the thermal conductivity of the final product. Replacing the natural aggregate with the expanded perlite reduced the thermal conductivity of the mixtures as a result of the porous structure of the perlite [22-36]. Thermal insulation materials in building envelopes are essential to improve thermal insulation performance and reduce energy consumption. Traditional thermal insulation materials used in building walls include organic thermal insulation materials with low thermal conductivity but poor fireproof performance and durability (expanded polystyrene (EPS), extruded polystyrene (XPS), and polyurethane (PUR)). Also, traditional thermal insulation includes inorganic materials such as rock wool, expanded perlite, and vermiculite, with good fireproof performance and durability but with higher thermal conductivity [55-58]. The thermal conductivity of perlite influences the reduction of the thermal conductivity of perlite concretes [31]. The reduction increases with the increases of the expanded perlite content.

A study made by Oktay et al. [50] reported that the use of perlite as a partial substitute for natural sand in concrete reduced thermal conductivity. Specifically, the study reported reductions of 22.96%, 38.01%, 64.23%, 74.36%, and 81.48% in 28-day thermal conductivity of concretes containing 10%, 20%, 30%, 40%, and 50% volume of perlite particles with the size range of 0.15-11mm respectively, as a substitution for natural sand. Using the different sizes of perlite particles can influence thermal conductivity, as shown in the study made by Jedidi et al. [31]. By partially substituting natural sand, by volume, with perlite particles with a size range of 2–4 mm, at levels of 15%, 30%, 45%, 60%, and 80%, it reduces up to 9.68%, 17.74%, 43.55%, 66.13%, and 79.03% in the 28-day thermal conductivity of perlite concrete [31].

In addition to its thermal insulation properties, perlite in concrete mixtures also improves resistance to freeze-thaw cycles and fire and provides improved acoustic comfort [55, 59, 60]. The unlinked open pores of the perlite form a cavity resonance sound absorption structure.

4. Conclusions and scope for future research

This paper aims to describe the state of knowledge and to review the previous studies that describe the properties of perlite concrete as a building material.

4.1. Conclusions

The use of perlite in construction materials has been the subject of extensive research over the past few decades. Perlite is a naturally occurring volcanic glass known for its lightweight and porous structure. Perlite is typically used as a lightweight aggregate in concrete, plaster, and other building materials in the construction industry. Through a comprehensive examination of previous studies, it is evident that perlite has the potential to be used as a component of building materials, mainly when mixed with concrete. The use of perlite in concrete can offer several advantages, including increased workability, reduced unit weight, improved fire resistance, and thermal and sound insulation. Its use in construction materials can reduce the density of the material, making it ideal for applications that require high levels of thermal and acoustic insulation.

Additionally, replacing natural aggregates as limestone and sand, with perlite helps reduce the consumption of natural resources. This can mitigate the negative impact of construction on the environment and promote sustainability in the construction industry.

However, it is also important to note that using perlite as an aggregate can negatively impact the mechanical properties of the concrete, including decreased compressive strength and increased water absorption. One of the most significant drawbacks is the reduction in compressive strength observed when perlite is used as an aggregate in concrete. This reduction in compressive strength can be attributed to the porous structure of perlite, which makes the concrete more prone to cracking and damage. Despite these limitations, perlite's porous structure can lower the material's density and improve its thermal and acoustic insulation properties.

The optimal proportion of perlite as an aggregate in concrete remains an area of ongoing research. Studies have suggested that a proportion of 5-30% of perlite as an aggregate slightly affects the compressive strength but significantly improves thermal conductivity and density. Adding different types of fibers or additives to improve mechanical properties should be an essential topic for future investigations. High-performance additives, such as silica fume, have also been explored to enhance perlite concrete's durability and mechanical properties.

Overall, the use of perlite in construction materials has the potential to offer significant benefits in various applications. However, the mechanical limitations of perlite concrete must be carefully considered when designing and implementing construction projects. Further research is needed to optimize the proportion of perlite in concrete and identify effective strategies to enhance the mechanical properties of perlite concrete. By addressing these challenges, using perlite in construction materials can be a promising approach to improve the sustainability and performance of building materials.

4.2. Future research

Perlite concrete has the potential to be a viable building material due to its numerous advantages, such as improved workability, reduced unit weight, improved fire resistance, and improved thermal and sound insulation. The properties of perlite can also reduce the consumption of natural resources by replacing traditional aggregates such as sand and limestone.

While the benefits of using perlite as an aggregate in concrete are numerous, the potential drawbacks also need to be considered. One of the main challenges of using perlite in concrete is that it can negatively impact the resulting mixture's mechanical and chemical properties. The porous structure of perlite can reduce the compressive strength of the concrete and increase its water absorption.

However, further studies are needed to analyze the effects of perlite as an aggregate in concrete regarding the mechanical and chemical properties of different mixtures. It is essential to conduct thorough laboratory testing to determine the optimal ratio of perlite for specific applications and to balance the properties of the concrete.

Moreover, it is crucial to balance the properties of perlite with the overall characteristics of the concrete, such as its curing time, setting time, and strength development. Different proportions of perlite can affect these properties. Therefore, a thorough analysis is necessary to ensure that the use of perlite does not compromise the quality and performance of the concrete.

Acknowledgments

This paper was elaborated with the support of the “Eco-innovative Products and Technologies for Energy Efficiency in Constructions – EFECON” research grant, project ID P_40_295/105524, Program co-financed by the European Regional Development Fund through Operational Program Competitiveness 2014-2020.

References

[1] Gatt D, Yousif C, Cellura M, Camilleri L and Guarino F 2020 Assessment of building energy modelling studies to meet the requirements of the new Energy Performance of Buildings Directive, *Renew. Sustain. Energy* Rev. **127**, 109886

[2] Arrigoni A, Panesar DK, Duhamel M, Opher T, Saxe S, Posen I D and MacLean H L 2020 Life cycle greenhouse gas emissions of concrete containing supplementary cementitious materials: cut-off vs. substitution, *J. Clean. Prod.* **263**, 121465

[3] Lehne J and Preston F 2018 Chatham house report making concrete change innovation in low-carbon cement and concrete the royal institute of international affairs, chatham house report series,www.chathamhouse.org/sites/default/files/publications/research/2018-06-13 (accessed on 02.02.2023)

[4] Snellings R 2016 Assessing, understanding and unlocking supplementary cementitious materials, *RILEM Tech.* Lett. **1**, 50

[5] Feifei F, Hanbin L, Zhong H and Hill A 2014 Development of a Carbon Emission Calculations System for Optimizing Building Plan Based on the LCA Framework. Hindawi Publishing Corporation, *Mathematical Problems in Engineering*

[6] Giama E and Papadopoulos AM 2020 Benchmarking carbon footprint and circularity in production processes: the case of stonewool and extruded polysterene, *J. Clean. Prod.* **257** 120559

[7] Santos SA, da Silva PR and de Brito J 2019 Durability evaluation of self-compacting concrete with recycled aggregates from the precast industry, *Mag. Concr. Res.* **71** (24), 1265–1282

[8] Gupta N, Siddique R and Belarbi R Sustainable and greener self-compacting concrete incorporating industrial by-products: a review, *J. Cleaner Prod.* **284**, 2021, 124803

[9] Barroqueiro T, da Silva PR and de Brito J 2020 High-performance self-compacting concrete with recycled aggregates from the precast industry: durability assessment, *Buildings* **10** (6) 113

[10] Abed M, Fořt J and Rashid K 2022 Multicriterial life cycle assessment of eco-efficient self-compacting concrete modified by waste perlite powder and/or recycled concrete aggregate, *Construction and Building Materials*, Volume **348**, 128696

[11] de Brito J and Kurda R 2021 The past and future of sustainable concrete: a critical review and new strategies on cement-based materials, J. Cleaner Prod. 281 **72**

[12] Park W J, Kim T, Roh S and Kim R 2019 Analysis of life cycle environmental impact of recycled aggregate, *Appl. Sci. Basel.* 9 (5) **13**

[13] Singh M and Garg M 1991 Perlite-based building materials – a review of current applications, *Constr. Build. Mater.* **5** (2), 75–81

[14] Rashad AM A 2016 synopsis about perlite as building material - a best practice guide for Civil Engineer, *Constr. Build. Mater.* **121**, 338–353

[15] Markiv T, Sobol K, Franus M and Franus W 2016 Mechanical and durability properties of concrete incorporating natural zeolite, *Arch. Civil Mech. Eng.* 16 (4), 554–562
[16] El Mir A 2018 Influence of Additives on the Porosity-Related Properties of Self Compacting concrete, Ph.D. thesis, Budapest University of Technology and Economics, Faculty of Civil Engineering
[17] Kimball Suzette M Menial 2016 Commodity Summaries, Department of the Interior, U.S. Geological Survey, 1–1202
[18] Bagdassarov N, Ritter F and Yanev Y 2001 Kinetics of perlite glasses degassing. TG and DSC analysis, *Fuel Energy Abstr.* **42**, 58
[19] Tekin N, Kadinci E, Demirbas O, Alkan M, Kara A and Doğan M 2006 Surface properties of poly(vinylimidazole)-adsorbed expanded perlite, *Microporous Mesoporous Mater.* **1**–3, 125–133
[20] Li R, Zhu J, Zhou W, Cheng X and Li Y 2016 Thermal compatibility of Sodium Nitrate/ Expanded Perlite composite phase change materials, *Appl. Therm. Eng*. 103, 452–458
[21] Sari A, Karaipekli A and Alkan C 2009 Preparation, characterization and thermal properties of lauric acid/expanded perlite as novel form-stable composite phase change material, *Chem. Eng. J.* **155**, 899–904
[22] Mladenovic A, Suput J S, Ducman V and Skapin AS 2004 Alkali-silica reactivity of some frequently used lightweight aggregates Cem Concr Res; **34**:1809–16
[23] Sodeyama K, Sakka Y and Kamino Y 1999 Preparation of fine expanded perlite, *J. Mater. Sci.* **34**, 2461–2468
[24] Demirboğa R and Gül R 2003 The effects of expanded perlite aggregate, silica fume and fly ash on the thermal conductivity of lightweight concrete, *Cem. Concr. Res.* **33**, 723–727.
[25] Topçu IB and Isıkdag˘ B 2008 Effect of expanded perlite aggregate on the properties of lightweight concrete, *J. Mater. Process. Technol.* **204**, 34–38
[26] Sengul O, Zaizi S, Karaosmanogu F and Tasdemir MA 2011 Effect of expanded perlite on the mechanical properties and thermal conductivity of lightweight concrete, *Energy Build.* **43**, 671–676
[27] Kramar D and Bindiganavile V 2013 Impact response of lightweight mortars containing expanded perlite, *Cement Concr. Compos.* **37**, 205–214.
[28] Lanzón M and García-Ruiz PA 2008 Lightweight cement mortars: Advantages and inconveniences of expanded perlite and its influence on fresh and hardened state and durability, *Constr. Build. Mater.* **22**, 1798–1806
[29] Silva LM, Ribeiro RA, Labrincha JA and Ferreira VM 2010 Role of lightweight fillers on the properties of a mixed-binder mortar, *Cement Concr. Compos*. **32**, 19–24
[30] Oktay H, Yumrutas R and Akpolat A 2015 Mechanical and thermophysical properties of lightweight aggregate concretes, *Constr. Build. Mater.* **96** 217–225
[31] Jedidi M, Benjeddou O and Soussi Ch 2015 Effect of Expanded Perlite Aggregate Dosage on Properties of Lightweight Concrete, *Jordan J. Civ. Eng.* **9**, 278– 291
[32] Demirboğa R and Gül R 2003 Thermal conductivity and compressive strength of expanded perlite aggregate concrete with mineral admixtures, *Energy Build.* **35**, 2003, 1155–1159
[33] Polat R, Demirboğa R and Khushefati WH 2015 Effects of nano and micro size of CaO and MgO, nano clay and expanded perlite aggregate on the autogenous shrinkage of mortar, *Constr. Build. Mater.* **81**, 268–275
[34] Sun D and Wang L 2015 Utilization of paraffin/expanded perlite materials to improve mechanical and thermal properties of cement mortar, Constr. Build. Mater. **101**, 791–796
[35] Isıkdag˘ B 2015 Characterization of lightweight ferrocement panels containing expanded perlite based mortar, *Constr. Build. Mater.* **8**, 15–23
[36] Yoon J Y, Kim J H, Hwang YY and Shin D K 2015 Lightweight concrete produced using a two-stage casting process, *Materials* (Basel) **8** (4), 1384–1397
[37] Yu L H, Ou H and Lee LL 2003 Investigation on pozzolanic effect of perlite powder in concrete,

Cem. Concr. Res. **33**, 73–76
[38] Szabo R, Dolgos F, Debreczeni A and G. Mucsi G 2022 Characterization of mechanically activated fly ash-based lightweight geopolymer composite prepared with ultrahigh expanded perlite content, *Ceram. Int.* **48**, 4261–4269
[39] Nematollahi B, Ranade R, Sanjayan J and Ramakrishnan S 2017 Thermal and mechanical properties of sustainable lightweight strain hardening geopolymer composites, Arch. Civ. Mech. Eng. **17**, 55–64
[40] Top S, Vapur H, Altiner M, Kaya D and Ekicibil A 2020 Properties of fly ash-based lightweight geopolymer concrete prepared using pumice and expanded perlite as aggregates, *J. Mol. Struct.* 1202, 127236
[41] Hamidi F, Valizadeh A and Aslani F 2022 The effect of scoria, perlite and crumb rubber aggregates on the fresh and mechanical properties of geopolymer concrete, *Structures*, **38**, 895-909
[42] Kapeluszna E, Kotwica L, Pichór W and Nocuń-Wczelik W 2020 Cement-based composites with waste expanded perlite - Structure, mechanical properties and durability in chloride and sulphate environments, Sustainable Materials and Technologies, **24**, 00160
[43] Adhikary SK, Ashish DK, Sharma H, Patel J, Rudžionis Z, Al-Ajamee M, Thomas BS and Khatib JM Lightweight self-compacting concrete: A review, Resources, *Conservation & Recycling Advances*, **15**, 2022, 200107
[44] Junaid MF, ur Rehman Z, Kuruc M, Medveď I, Bačinskas D, Čurpek J, Čekon M, Ijaz N and Ansari WS 2022 Lightweight concrete from a perspective of sustainable reuse of waste byproducts, *Construction and Building Materials*, **319**, 126061
[45] Alyousef R, Benjeddou O, Soussi C, Khadimallah M A and Jedidi M 2011 Experimental study of new insulation lightweight concrete block floor based on perlite aggregate, natural sand, and sand obtained from marble waste, *Adv. Mater. Sci. Eng.*
[46] Balta I and Cilasun N 1993 Perlite concrete, Developments in Civil Engineering, Technical Congress, **1**, pp. 572–579
[47] Jiang L, Lu W, Wang W, Zhang Y, Han Q and Li Z 2022 Mechanical properties and frost resistance of self-healing concrete based on expended perlite immobilized bacteria, Construction and Building Materials, **348**, 128652
[48] Wang X, Wu D, Geng Q, Hou D, Wang M, Li L, Wang P, Chen D and Sun Z 2021 Characterization of sustainable ultra-high performance concrete (UHPC) including expanded perlite, *Construction and Building Materials* **303**, 124245
[49] Karakoç MB and Denirboga R 2010 HSC with expanded perlite aggregate at wet and dry curing conditions, *J. Mater. Civil Eng.* (ASCE) **22** (12), 1251–1259
[50] Oktay H, Yumruts R and Akpolat A 2015 Mechanical and thermophysical properties of lightweigh aggregate concretes, *Constr. Build. Mater*. **96**, 217–225
[51] Sriwattanapong M, Sinsiri T, Pantawee S and Chindaprasirt P 2013 A study of lightweight concrete admixed with perlite, Suranarre J. *Sci. Technol.* **20** (3), 227–234
[52] Türkmen I and Kantarci A 2006 Effects of expanded perlite aggregate and different curing conditions on the drying shrinkage of self-compacting concrete, *Indian J. Eng. Mater. Sci.* **13**, 247–252
[53] Türkmen I and Kantarci A 2007 Effects of expanded perlite aggregate and different curing conditions on the physical and mechanical properties of self-compacting concrete, *Build. Environ.* **42**, 2378–2383
[54] Panagiotopoulou Ch, Angelopoulos PM, Kosmidi D, Angelou I, Sakellariou L and Taxiarchou M 2022 Study of the influence of the addition of closed- structure expanded perlite microspheres on the density and compressive strength of cement pastes, Materials Today: Proceedings, **54**, Part 1, 118-124
[55] Jia G and Li Z 2021 Influence of the aerogel/expanded perlite composite as thermal insulation aggregate on the cement-based materials: Preparation, property, and microstructure *Constr.*

Build. Mater., **273**, Article 121728

[56] Papadopoulos A M 2005 State of the art in thermal insulation materials and aims for future developments, *Energy Build.* **37** (1), 77—86

[57] Jiang L, Xiao H, An W, Zhou Y and Sun J 2014 Correlation study between flammability and the width of organic thermal insulation materials for building exterior walls, *Energy Build.* **82**, 243—249

[58] Yin X, Li H, Bo H and Ke H 2010 Weatherability studies on external insulation thermal system of expanded polystyrene board, polystyrene granule and polyurethane foam, J. Wuhan Univ. Technol.-*Mat. Sci. Edit.* **25** (6), 1027—1032

[59] Mo X and Fournier B 2007 Investigation of structural properties associated with alkali-silica reaction by means of macro- and micro-structural analysis Mater. Charact., **58** (2) 179-189

[60] Abdul Ajeej K, Varun Teja K and Meena T 2019 Experimental study on mechanical properties and micro-structure of perlite powder concrete subjected to elevated temperatures *Int. J. Recent Technol. Eng.,* **7** (6), pp. 1350-1357

Heavy and light masonry walls impact on medium height reinforced concrete frames buildings

S Constantinescu[1*]

[1] Technical University of Bucharest, Romania

* sorina.constantinescu@yahoo.com

Abstract. The paper studies the influence of light weight aerated autoclaved concrete (AAC) partitioning walls to a medium height reinforced concrete structure, in Bucharest (Romania). The building is designed as a framed structure with non-bearing AAC walls placed in certain bays. There is a bearing full bricks masonry wall in one of the bays. Full bricks are used because of its high strength compared to AAC. The aim is to determine the importance of the masonry walls placement, both full bricks and AAC, for the building's behavior. The analysis points out the structure's elastic and plastic behavior and also the efforts developed in the non-bearing walls. The bearing masonry wall's behavior is very important, because the structure is built in a high seismic area. To extend the study, the full bricks wall is replaced with a reinforced concrete one. This is done to see how it impacts the building's behavior, for both the elastic and plastic stage. The paper concludes that AAC, full bricks and concrete walls placements have a significant influence on the structure.

1. Introduction

Masonry is one of the oldest building materials and it is still used today. AAC can be used for bearing and nonbearing walls. In low seismic areas, such as one or two story buildings, the damage limit may not be reached. In high seismic areas, significant damage may be expected. This was concluded from tests [1, 2]. For framed buildings without nonbearing walls, the beams and columns bending assures the resistance to seismic forces. The real strength for buildings can be evaluated by nonlinear pushover analysis [3]. Masonry infill walls increase strength, stiffness and energy dissipation for buildings. Such buildings perform better in moderate earthquakes even if they were not designed to withstand such seismic forces [4].

Tests show, infill walls can provide alternative paths for transferring loads initially taken by the beams, so they increase the frames strength [5]. Masonry infill walls with rigid connections to the structure cause undesired effects during seismic events, like short-column effect or soft story effect, even torsion and out of plane collapse. Also, in frames, it was seen that plastic hinges appear at the columns ends if the weak beams and strong columns rule isn't followed. Cast in place slabs increase the beams stiffness significantly [6]. Masonry walls don't behave well when subjected to eccentric loads, as they crack at 40 to 60% of the maximum load [7]. According to nonlinear static analysis, confined masonry walls have important advantages in comparison to the unreinforced walls, in seismic areas [8]. Masonry walls reinforcements help them to work together with the confining elements [9].

If masonry walls continuous horizontal reinforcement is used, it helps the structure to reach a ductile behavior [10]. However, seismic shear failure is expected to occur in walls, for dual buildings. The frames usually show large deformations in the plastic stage [11]. The paper studies the difference in behavior for a reinforced concrete framed structure with a bearing full bricks (FB) masonry wall in 4 different bays. It will be used as an office building. It also contains partitioning aerated autoclaved (AAC) walls placed in some bays and lightweight partitioning walls in other bays placed away from the bearing wall. The structure in study is of medium height, situated in a high seismic area (Bucharest, Romania). The structure's behavior is analyzed for each placement of the masonry bearing wall, in both elastic and plastic stage. To extend the study further, the masonry FB wall is replaced with a reinforced concrete one (RC), and the building's behavior is analyzed again. The bearing and nonbearing walls efforts and stresses are also shown. It is important to highlight if the masonry FB wall can bare the efforts it is subjected to when it is placed in different bays. The codes in force used to design the structure are [12-18].

2. Building description

The building in study shows a reinforced concrete framed structure with a full bricks (FB) confined masonry bearing wall placed in one bay.

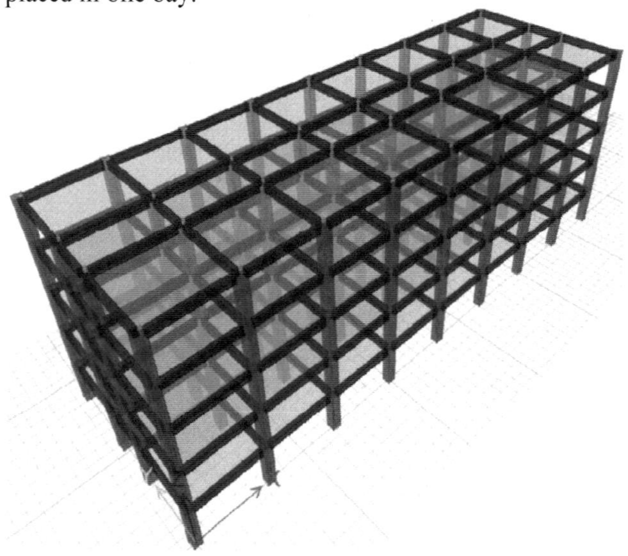

Figure 1. 3D frames building.

There are also some partitioning AAC walls in different bays, inserted in the model as walls and some lightweight walls in other bays, inserted in the model as loads. All the AAC walls rest on beams. Figure 1 shows the 3D image of the building without any walls to be more general. Figures 2 to 5 show the current building plan for the 4 cases studied with the FB/RC (P1) and the AAC walls (P2 to P13). There are 4 cases studied here (A FB to D FB) if the FB is present and 4 (A RC to D RC), if the RC is present. The bays on direction X are 5 m long, while these on Y measure 5 m, 3 m and 5 m.

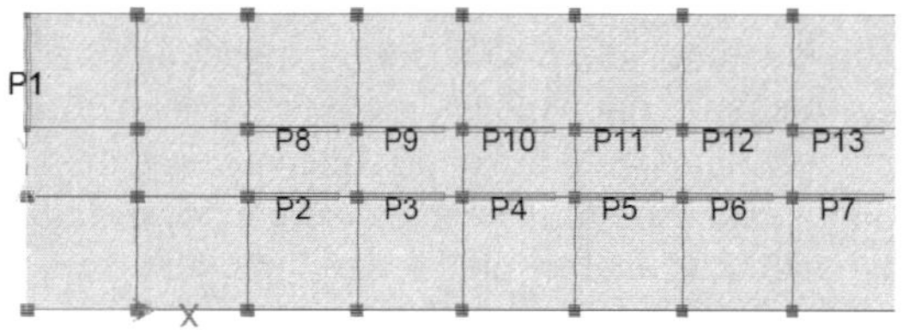

Figure 2. Floor plan for cases A FB/RC.

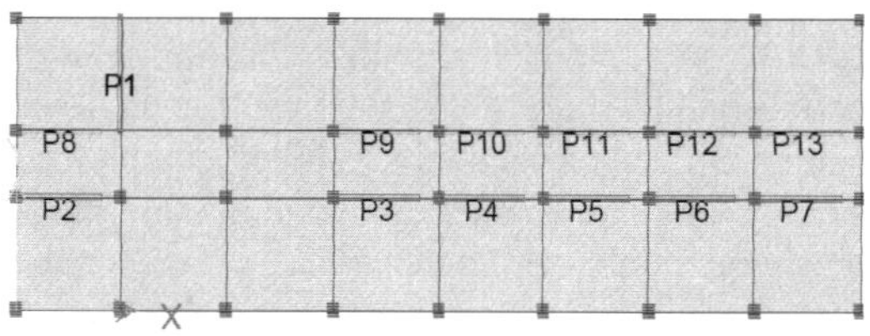

Figure 3. Floor plan for cases B FB/RC.

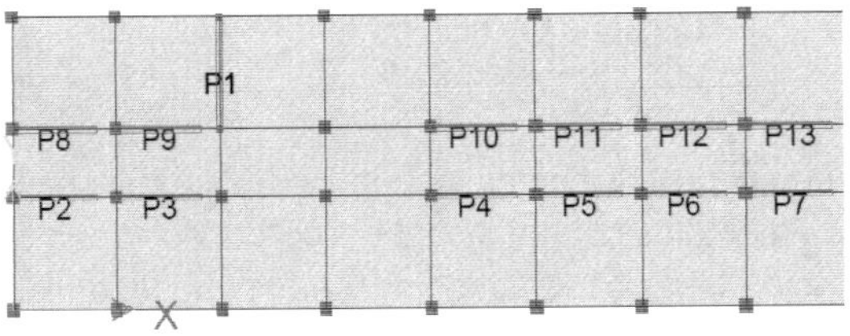

Figure 4. Floor plan for cases C FB/RC.

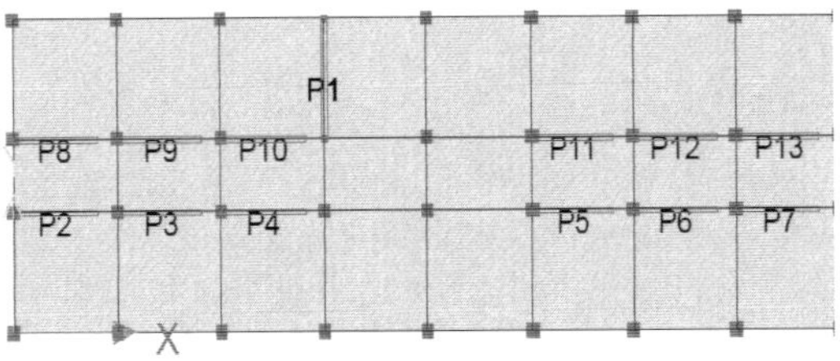

Figure 5. Floor plan for cases D FB/RC.

3. Theory elements used in the paper

3.1. Materials design strengths

The concrete used is C20/25, the full bricks elasticity modulus is E = 4400 N/mm^2, with standard strength f_b = 10 N/mm^2 and M10, for AAC blocks E = 3424 N/mm^2, f_b = 7 N/mm^2 and M10, steel reinforcement bars are S500 [14] and [18]. The stresses analysed for AAC are σ_x, σ_z and τ_{xz}. They are compared to the design strengths horizontal compression f_{dh}, vertical compression f_d, shear strength for horizontal direction $f_{vd,0}$. The horizontal and vertical strengths perpendicular to the wall f_{xd1} and f_{xd2} are used to calculate the bearing bending moments perpendicular to the wall plane M_{Rxd1} and M_{Rxd2} [18]. f_{dh} = 1.1 N/mm^2, f_d = 2.23 N/mm^2, $f_{vd,0}$ = 0.13 N/mm^2, f_{xd1} = 0.05 N/mm^2 and f_{xd2} = 0.10 N/mm^2 for the AAC walls. The design strengths are determined from their characteristic values: f_{kh}, f_k, $f_{vk,0}$, f_{xk1} and f_{xk2}. The masonry strengths insurance factor is γ_M = 1.9 [15] and [18]. f_d = 2.31 N/mm^2 and $f_{vd,0}$ = 0.158 N/mm^2 are the strengths for the FB wall. The concrete compression design strength f_{cd} in (8) is determined using the characteristic strength f_{ck}. For rebars, f_{yd} = 435 N/mm^2 for S500 and f_{yd} = 308 N/mm^2 for S355, while f_{yk} is the characteristic value. γ_C = 1.5, for concrete and γ_S = 1.15 for steel [14]. Equation (1) is used for all materials.

$$f_d = f_k/\gamma_{M\ (C,\ S)}\ [\mathrm{N/mm^2}] \tag{1}$$

3.2. Seismic load

The seismic load is introduced by coefficient c_s. The base force F_b in equation (2) is calculated using [16-17]. The building's earthquake importance-exposure coefficient $\gamma_{I,e}$ = 1, the maximum value in the elastic spectrum β_0 = 2.5, q is the structure's behavior factor, q = 3.5·1.35 for medium ductility structures DCM [17], m is the building's mass, reduction factor η = 1, λ = 0.85 and a_g = 0.30g [17].

$$F_b = \gamma_{I,e} \cdot \beta_0 \cdot a_g/q \cdot m \cdot \eta \cdot \lambda = c_s \cdot m \cdot g = 0.14 \cdot m \cdot g\ [\mathrm{kN}] \tag{2}$$

3.3. Masonry walls bearing efforts parallel to the walls plane

M_{Rd} in (3) is the wall's bearing bending moment [18]. $M_{Rd(M)}$ in (5) and $M_{Rd(As)}$ in (6) are the bearing bending moments from the masonry area and from the slender columns reinforcements respectively [18]. A1 is the wall's compressed area [18].

$$M_{Rd} = M_{Rd(M)} + M_{Rd(As)}\ [\mathrm{kNm}] \tag{3}$$

$$A1 = N_{Ed}/(0.85 \cdot f_d)\ [\mathrm{mm2}] \tag{4}$$

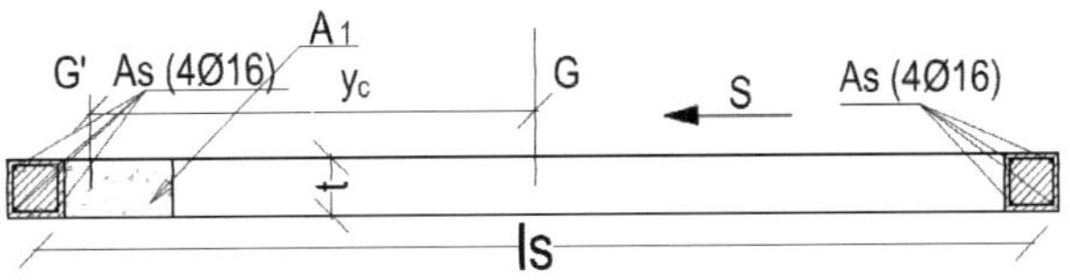

Figure 6. Confined masonry wall section with reinforcement bars.

$$M_{Rd(M)} = N_{Ed} \cdot yc \text{ [kNm]} \quad (5)$$

$$M_{Rd(As)} = l_s \cdot A_s \cdot f_{yd} \text{ [kNm]} \quad (6)$$

In figure 6, y_c is the distance between the compressed masonry area weight centre G' and the wall's weight centre G [18]. ls is the distance between the slender columns at the wall's edges. As is the slender columns horizontal reinforcement area. t = 25cm is the walls width and S is the seismic direction. V_{Rd} is the masonry wall bearing shear force and V_{Ed} is the horizontal shear force from the seismic loads combination: permanent load +0.4 ·variable load + seismic load [12]. The fundamental loads combination: 1.35· permanent load +1.5 ·variable load [12], is also used for the masonry wall, to establish if it can bare the maximum axial load at its bottom story. h_{pan} = story height - 30cm (tie beam height) and l_{pan} = ls - t are the height and length of the masonry area panel. V_{Rd2} in (10) is the bearing horizontal shear force from the reinforcement in the slender column at walls compressed edge [18]. As is the reinforcement area in the slender column at the walls compressed edge. λc is the reinforcement participation factor. Here, λc = 0.20, for longitudinal reinforcement Φ16. V_{RSC} in (11) is the slender columns shearing capacity at the walls compressed area.

$$V_{Rd} = \text{minimum } (V_{Rd1}*+, V_{Rd,i}*) + V_{Rd2} + V_{RSC} \text{ [kN]} \quad (7)$$

$$V_{Rd1}* = 0.4 \bullet (N_{Ed} + 0.8 \cdot V_{Ed} \cdot h_{pan}/l_{pan}) \text{[kN]} \quad (8)$$

$$V_{Ed} \leq l_{pan} \cdot t \cdot f_{vd,0} \text{ [kN]} \quad (9)$$

$$V_{Rd2} = \lambda c \cdot As \cdot f_{yd} \text{ [kN]} \quad (10)$$

$$V_{RSC} = Ac \cdot f_{cvd} \text{ [kN]} \quad (11)$$

In equation (15), b is a correction coefficient connected to the masonry walls dimensions h_{pan} = h = story height - beam height and $l_{pan} = l_w$ = ls - t are the height and length of the masonry area panel. $V_{Rd,1}*$ and $V_{Rd,i}*$ are the bearing shear forces on the horizontal and diagonal directions, b = 1.5 if h/lw ≥ 1.5, b = 1.0 if h/lw < 1.0 and b = h/lw if 1.0 ≤ h/lw < 1.5. $\sigma_{0d} = N_{Ed}/(t \cdot lw)$ [18].

$$f_{vd,i} = f_{vk,i}/1.9 = 0.22 \cdot f_{bt}(1+5*\sigma_{0d}/f_{bt})^{0.5}/1.9 \text{ [N/mm}^2\text{]} \quad (12)$$

$$V_{Rd,1}* = 0.4 \cdot (N_{Ed} + 0.8 \cdot V_{Ed} \cdot h_{pan}/l_{pan}) \text{ [kN]} \quad (13)$$

$$V_{Ed} \leq lpan \cdot t \cdot f_{vd,0} \quad (14)$$

$$V_{Rd,i} = Aw \cdot f_{vd,i} / b \quad (15)$$

N_{Rd} is the masonry wall's bearing axial force. In formula (16), $\Phi_{i(m)}$ = buckling coefficient according to the wall's slenderness and eccentric loads and A is the wall's section area. $\Phi_{i(m)}$ = minimum (Φ_i (for top and bottom) and Φ_m (for the midsection)). In equation (17), t is the wall's thickness, e_i is the eccentricity from the wall's centre plane, calculated using 3 other eccentricities e_{0i} (for vertical loads), e_{hi} (from bending moments perpendicular to the wall $M_{hm(i)}$) and e_a=maximum (t/30, story height/300, 1cm) [18]. N_1 is the axial force from the wall at the above story, N_2 is from the slabs loading the wall in study, d_1 and d_2 are the eccentricities respectively. Φ_m will be taken from table 1, according to e_m/t, where e_m is the eccentricity at the walls midsection. AAC walls here do not have any slender columns included, while FB walls do. This is why there are different values for Φ_m [18].

$$N_{Rd} = \Phi_{i(m)} \cdot A* f_d \quad (16)$$

$$\Phi_i = 1-2*ei/t \quad (17)$$

$$e_i = e_{0i} + e_{hi} + e_a \geq 0.05*t \quad (18)$$

$$e_{0i} = (N_1*d_1 + \Sigma N_2*d_2)/(N_1 + \Sigma N_2) \quad (19)$$

$$e_{hm(i)} = M_{hm(i)}/(N_1 + \Sigma N_2) \quad (20)$$

$$e_m = 2/3 \bullet e_{0i} + e_{hm} + e_a \quad (21)$$

Table 1. Buckling coefficient for the wall's midsection.

Slenderness (story height/t)	Walls	0.05	0.10	0.15	0.20	0.25	0.30
Maximum 12	AAC	0.80	0.70	0.59	0.49	0.38	0.28
Maximum 15	FB	0.75	0.64	0.53	0.42	0.32	0.22

3.4. Masonry walls bearing efforts perpendicular to the walls plane

The bearing bending moments perpendicular to the masonry walls M_{Rxd1} in equation (22) (horizontal) and M_{Rxd2} in equation (23) (vertical) values will be compared to the design bending moments values, M_{Exd1} and M_{Exd2} [12] calculated from the model respectively. $W_w = 1000 \cdot t^2/6$ is the wall resistance modulus (in mm^3/m), t is the wall thickness, σ_{dw} is the compression stress at the wall's middle height section [12], γ_{mas} AAC = 9kN/m^3 is the masonry weight per cubic meter and Hw is masonry wall height.

$$M_{Rxd1} = W_w \cdot (f_{xd1} + \sigma_{dw}) \text{ [kNm/m]} \quad (22)$$

$$M_{Rxd2} = W_w \cdot f_{xd2} \text{ [kNm/m]} \quad (23)$$

$$\sigma_{dw} = \gamma_{mas} \cdot Hw/2 \text{ [N/mm}^2\text{]} \quad (24)$$

$$W_w = 1000 \cdot t^2/6 = 1000 \cdot 2502/6 = 10416666 \text{ mm}^3\text{/m} \quad (25)$$

4. Elastic analysis results

Dimensions and longitudinal reinforcement in beams and tie beams in the FB wall is 3Φ18 at the section top and bottom, 3Φ22 on each side for columns, 2Φ16 on each side in slender columns.

4.1. Bearing masonry wall results

Effective efforts N_{Ed}, M_{Ed} and V_{Ed} and bearing efforts M_{Rd} and V_{Rd} in the FB wall are written for all stories. They are calculated for the seismic loads combination parallel to wall's direction. Axial efforts $N_{Ed\ G}$ and $N_{Rd\ G}$ for the fundamental loads combination are written only for the ground stories, where they reach higher values. All efforts are presented for all 4 cases in which the bearing masonry wall is used. Each case is displayed separately in table 2.

In table 2, it is seen that the effective efforts are below the bearing ones, except for 2 situations. For cases C and D it is seen that the effective bending moments M_{Ed} at story 1 surpass the bearing bending moments M_{Rd}. It means the FB bearing wall does not withstand the efforts it would be subjected to in cases mentioned above. This situation can be resolved by increasing the slender columns areas and reinforcements. However, this solution may lead to large concrete elements that might not work well with the masonry panels. This can be included in further studies to see if it is appropriate or not.

4.2. AAC walls results

Results for the AAC walls, calculated for the seismic loads combination parallel to the walls plane are: N_{Ed} = 28 kN, M_{Ed} = 10 kNm, M_{Rd} = 56 kNm, V_{Ed} = 3 kN and V_{Rd} = 11 kN. It is important to mention that M_{Ed} and V_{Ed} are from the partitioning walls at the top story, where they are higher. This is because the seismic force is higher there.

The same thing can be said about the bending moments perpendicular to the walls plane M_{Exd1} and M_{Rxd1} for the horizontal direction and M_{Exd2} and M_{Rxd2} for the vertical. M_{Rxd1} is 0.64 kNm/m, and M_{Rxd2} = 1.04 kNm/m, both being surpassed by effective values at the walls corners, as figures 7 and 8 show. This suggests cracks may appear there. For the AAC walls, N_{Ed} = 37 kN in the fundamental combination, while N_{Rd} = 1845 kN. Here, AAC walls are used for partitioning. They can also be used as bearing elements, so all their efforts would be significantly greater. Figures 9 to 20 show stress values in AAC walls in the seismic loads combination on direction X, in N/mm^2.

Table 2. Efforts in FB bearing wall.

	N_{Ed} (kN)	M_{Ed} (kNm)	M_{Rd} (kNm)	V_{Ed} (kN)	V_{Rd} (kN)	$N_{Ed\ G}$ (kN)	$N_{Rd\ G}$ (kN)
				Case A			
Story 1	1332	3094	3770	455	635	1273	2210
Story 2	1066	2022	3715	457	529		
Story 3	789	991	2856	337	418		
Story 4	516	349	2496	206	309		
Story 5	249	215	638	45	177		
				Case B			
Story 1	2022	3552	3713	547	975	2596	2907
Story 2	1601	2336	3871	586	743		
Story 3	1179	1068	3321	447	574		
Story 4	769	614	3032	273	410		
Story 5	374	358	2152	60	229		
				Case C			
Story 1	2074	4101	3611	629	1005	2622	2908
Story 2	1645	2674	3879	681	761		
Story 3	1209	1173	3712	513	586		
Story 4	785	729	3070	297	417		
Story 5	377	380	2161	44	228		
				Case D			
Story 1	2108	4543	3641	696	1024	2623	2908
Story 2	1671	2971	3870	758	771		
Story 3	1226	1284	3793	567	593		
Story 4	794	804	2881	320	420		
Story 5	380	405	2184	39	288		

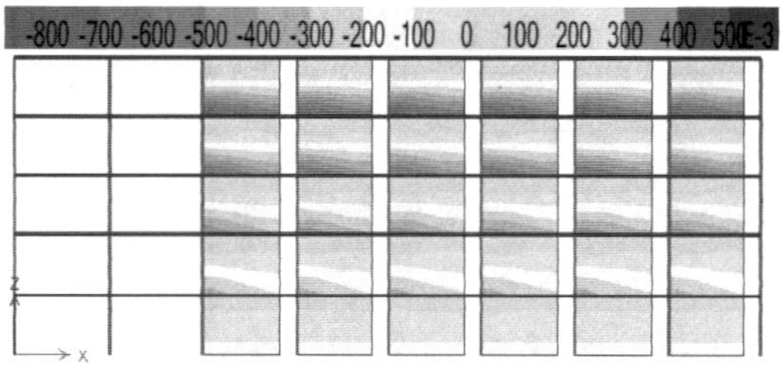

Figure 7. M_{Exd1} in AAC walls.

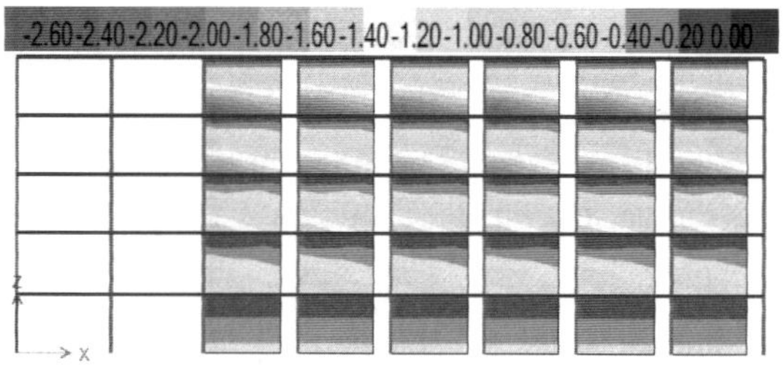

Figure 8. M_{Exd2} in AAC walls.

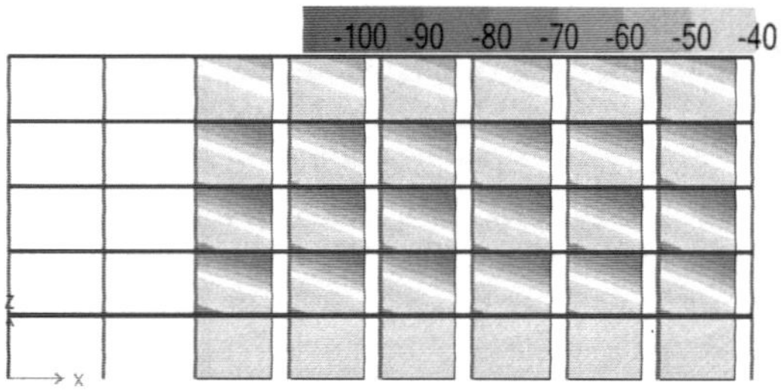

Figure 9. σ_x in case A FB.

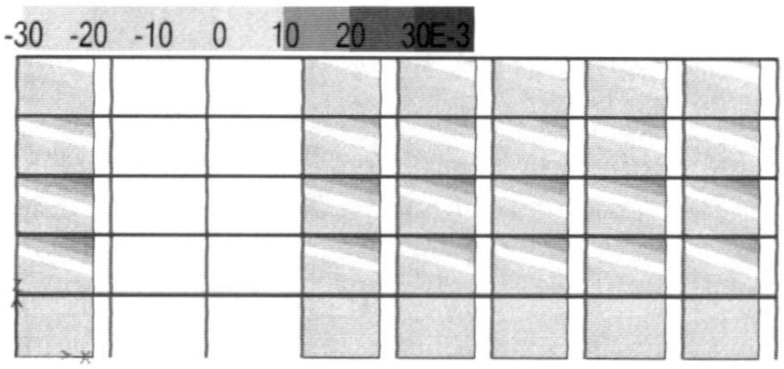

Figure 10. σ_x in case B FB.

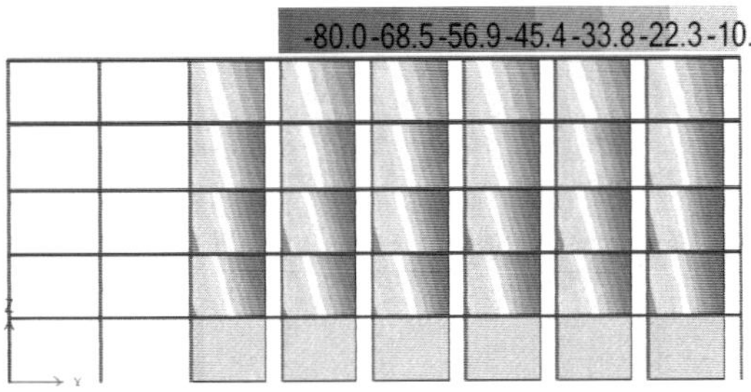

Figure 11. σ_z in case A FB.

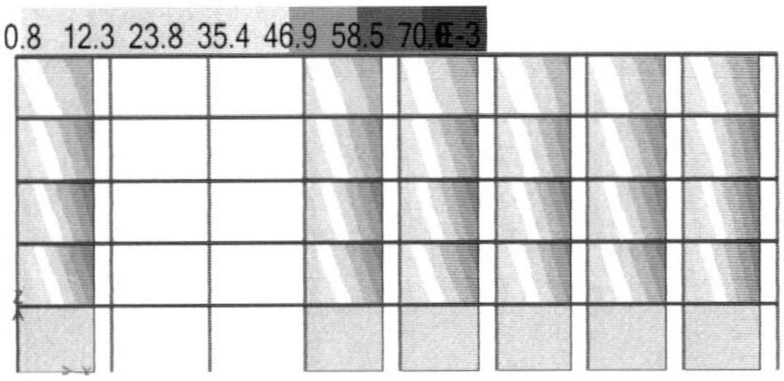

Figure 12. σ_z in case B FB.

All stresses show the walls compressed diagonally, however σ_z is developed more vertically. They are below the bearing strengths. It is noticed the values are higher for case A. This is explained by the further placement they have from the bearing wall that does assure some level of stiffness to the structure. This also happens in case D, but far less evident. It may be concluded that the FB wall creates a stiffer area around it and tends to take on some of the seismic loads that would otherwise be taken by the AAC walls. It can't do that very effectively if the AAC walls are too far from it, or if there are too many AAC walls too far from it, like in case D. It is also clear that stresses reach low values at ground floor story. This may be because they are connected to the embedded part of the structure.

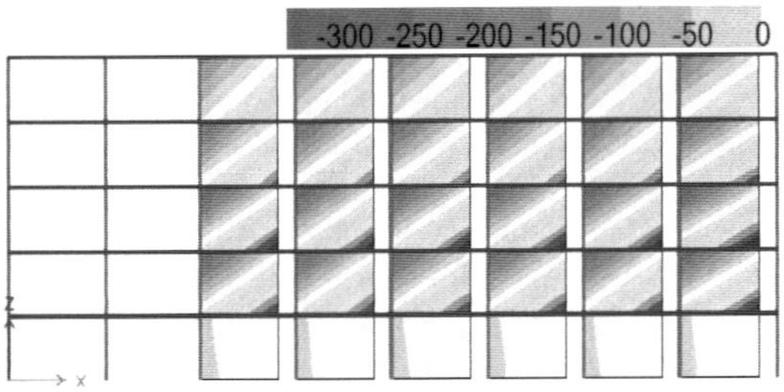

Figure 13. τ_{xz} in case A FB.

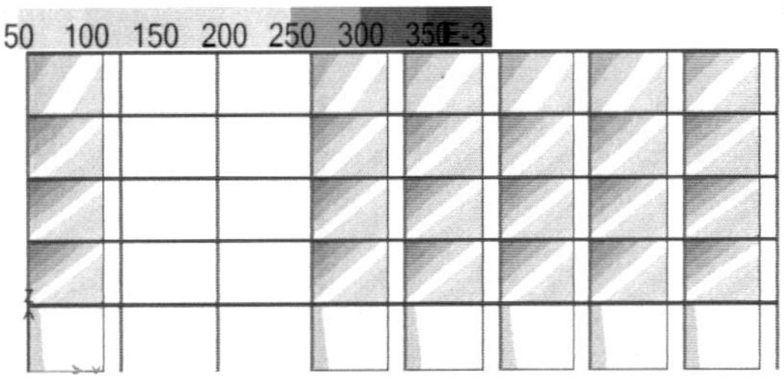

Figure 14. τ_{xz} in case B FB.

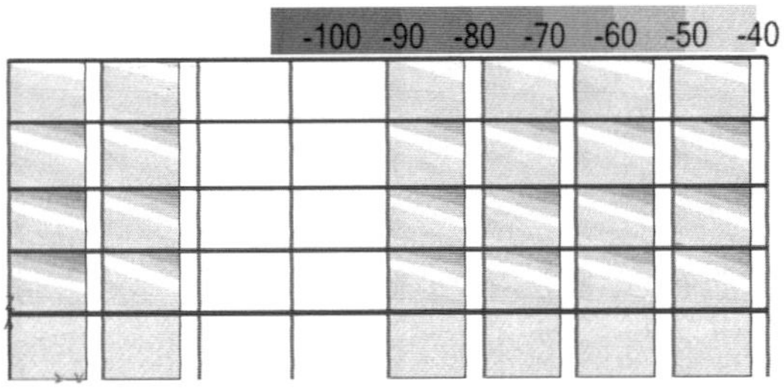

Figure 15. σ_x in case C FB.

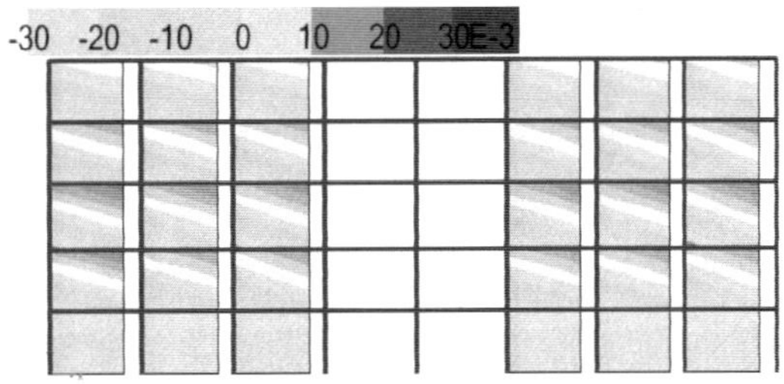

Figure 16. σ_x in case D FB.

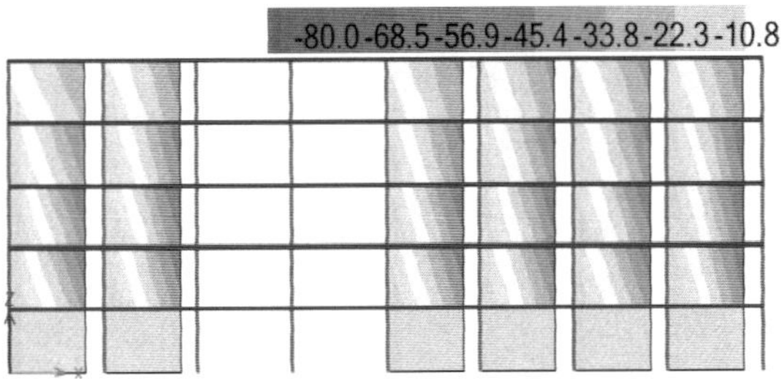

Figure 17. σ_z in case C FB.

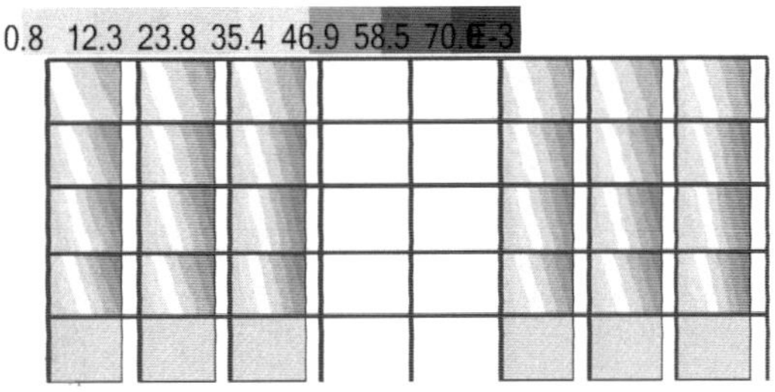

Figure 18. σ_z in case D FB.

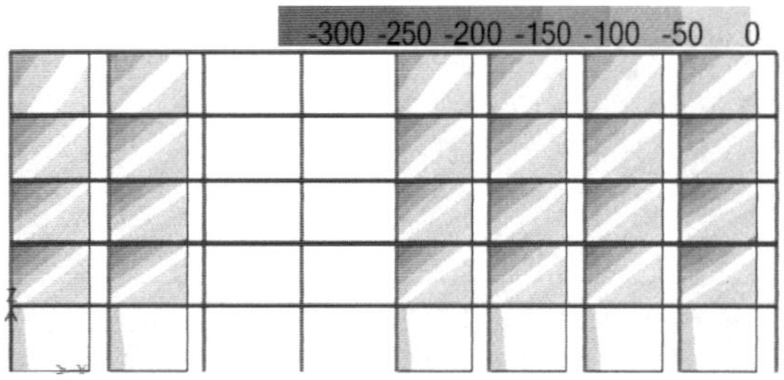

Figure 19. τ_{xz} in case C FB.

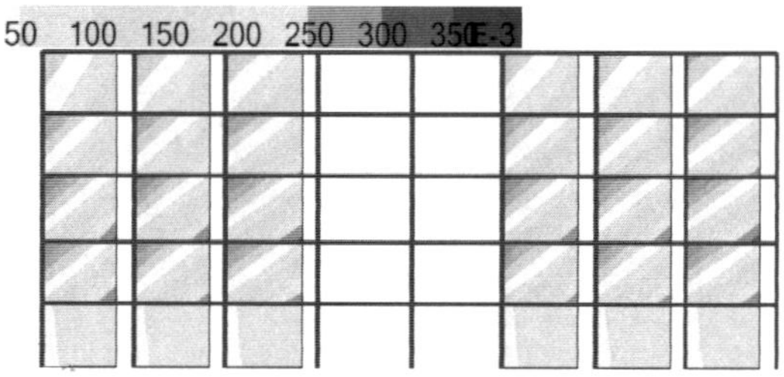

Figure 20. τ_{xz} in case D FB.

5. Plastic stage results

5.1. Plastic hinges development in beams

The hinges colour code is: green for the plastic hinges being formed (stage B), blue for hinges that have reached their limit capacity (C), pink means the load has been redistributed (D) and red is for collapse (E). If the FB wall is present (in figures 21 to 24), many plastic hinges are developed up to stage E, particularly in beams placed away from the AAC walls, for direction X, and away from the FB wall for direction Y. This is explained by the stiffness, these walls provide to the structure. This is seen clearer if the RC wall is present, because it is more rigid than the FB wall. This wall also influences the structure slightly on direction X, as it is seen in figures 25 to 28. It does not have such a strong effect if it is placed at the building edge. It is noticed, that both bearing walls tend to hold in place the area of the structure they are directly attached to, for direction Y.

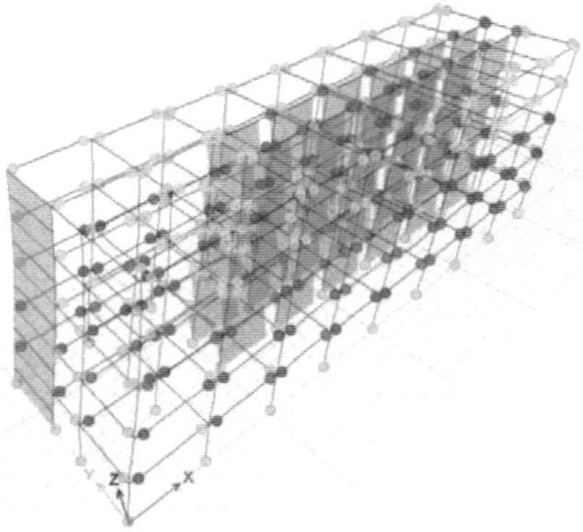

Figure 21. A FB on X.

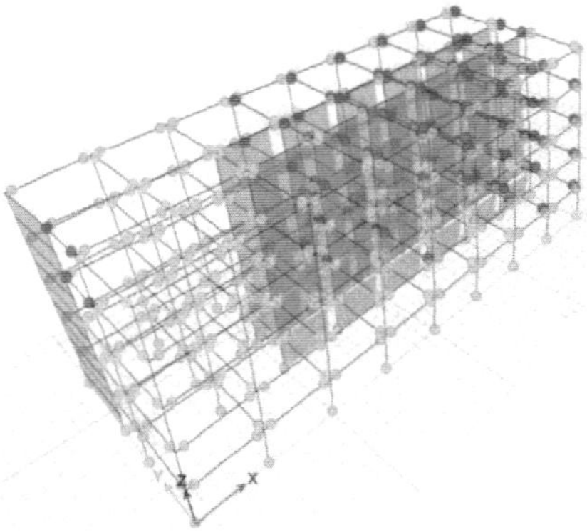

Figure 22. A FB on Y.

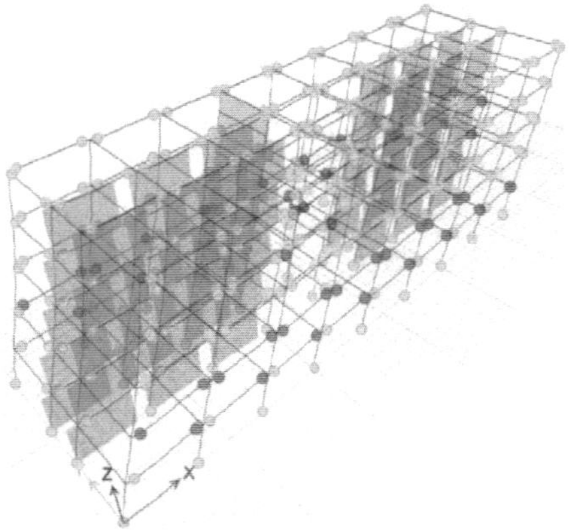

Figure 23. D FB on X.

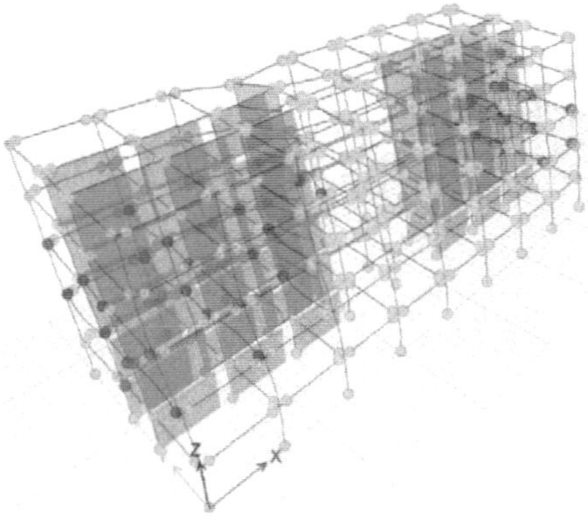

Figure 24. D FB on Y.

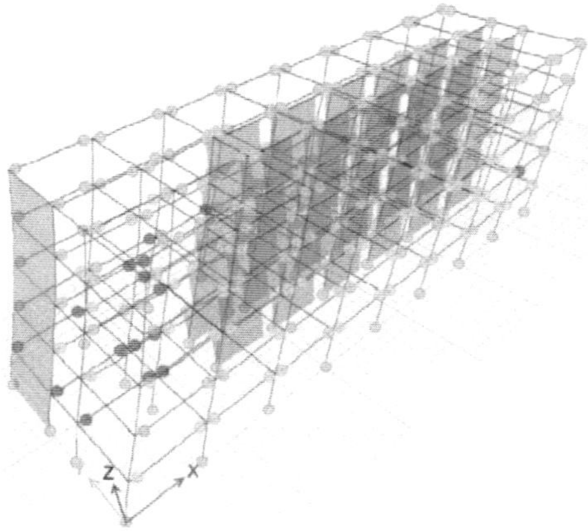

Figure 25. A RC on X.

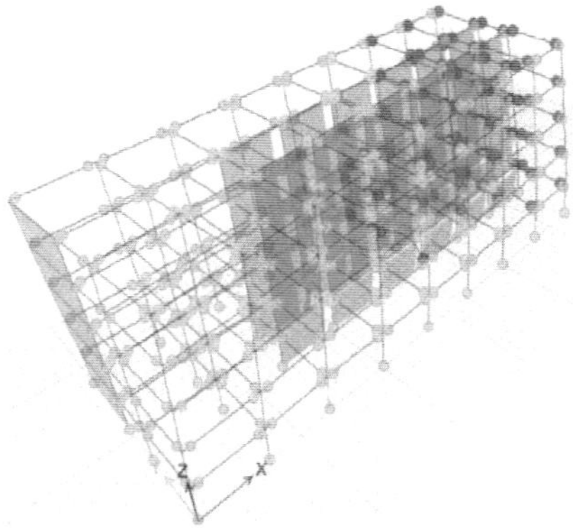

Figure 26. A RC on Y.

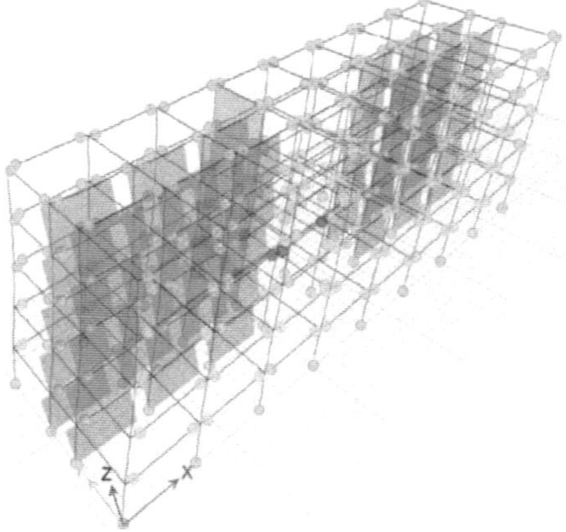

Figure 27. D RC on X.

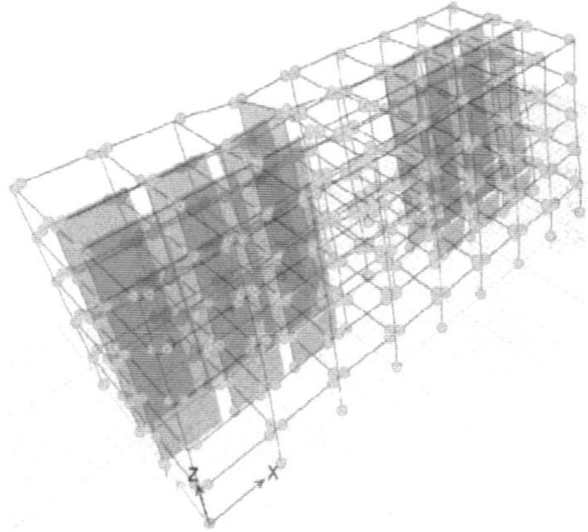

Figure 28. D RC on Y.

5.2. Pushover diagrams

The pushover diagrams for direction X, look similar if either the bearing FB or RC wall is present. For cases FB X and RC X, the AAC walls are absent. For cases A to D, with FB or RC, for directions X or Y, the AAC walls are present. The structure's rigidity (the diagrams slope) is the same for all cases, until the base force reaches 60 MN. It may be explained by the presence of the same AAC walls in both cases, and by the placing of a bearing wall (FB or RC) on direction Y, so the structure is less affected on direction X. If the AAC walls would not be present, the structure's stiffness significantly drops, as the dark blue lines show in cases FB X and RC X. On direction Y, the RC wall does have an important effect in stiffening the structure. If the FB wall is used, the diagrams show large top displacements and an important rigidity drop for case A. Rigidity is maintained mostly the same if the RC wall is there, but the top displacements and maximum base forces reached are lower compared to the FB cases (11cm compared to 41 cm and 101 MN compared to 134 MN, just to give an example). There are however some exceptions. For case A, when the bearing wall is at the edge, its effect is significantly reduced. For the FB wall case, diagram A shows rigidity drops and increases followed by a sharp drop at the analysis end. For the RC wall, case A also shows stiffness decreases and increases.

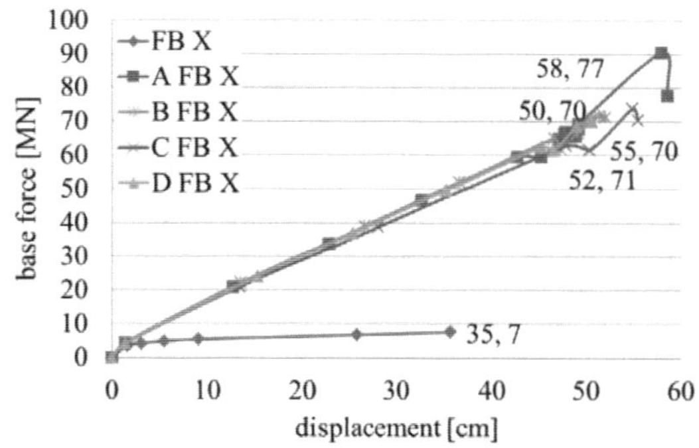

Figure 29. Pushover diagram for FB on X.

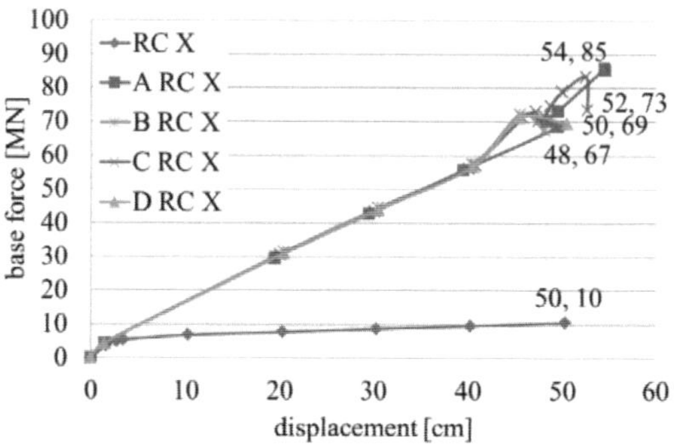

Figure 30. Pushover diagram for RC on X.

The conclusions for the pushover analysis on direction X are: the buildings with FB and RC walls behave similarly on direction X and the AAC walls increase the structures rigidity on X (the diagrams slopes are greater if they are present). The conclusions for direction Y are: the FB structure has larger displacements and base forces but cases A and D show lower structure stiffness. The top displacements and base forces values are presented in table 3. The diagrams are also important as they show a better comparison between the developments of these values for all cases.

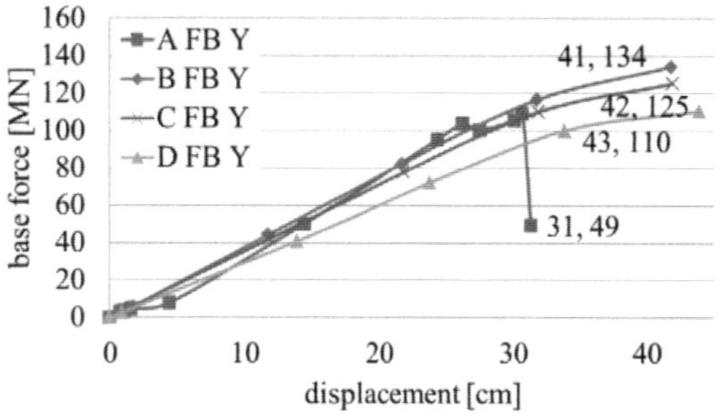

Figure 31. Pushover diagram for FB on Y.

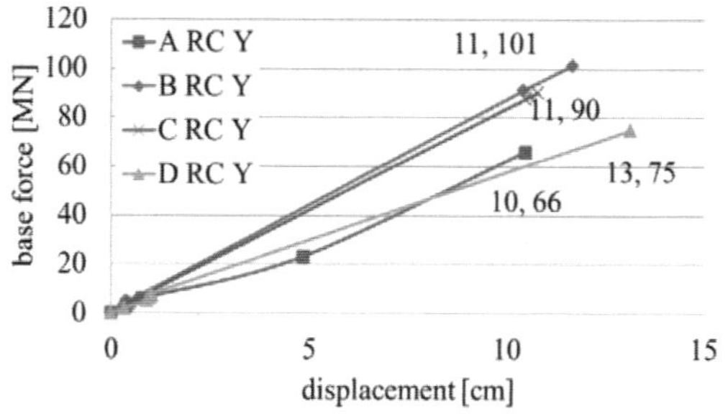

Figure 32. Pushover diagram for RC on Y.

Table 3. Pushover diagrams base forces and top displacements values.

Case	Top displacement (cm)	Base force (MN)	Case	Top displacement (cm)	Base force (MN)
Cases on direction X					
FB X	35	7	RC X	50	10
A FB X	58	77	A RC X	54	85
B FB X	52	71	B RC X	48	67
C FB X	55	70	C RC X	52	73
D FB X	50	70	D RC X	50	69
Cases on direction Y					
A FB Y	31	49	A RC Y	10	66
B FB Y	41	134	B RC Y	11	101
C FB Y	42	125	C RC Y	11	90
D FB Y	43	110	D RC Y	13	75

6. Conclusions

Both bearing and non-bearing masonry walls have a great stiffening effect on the structure for all cases studied here. The bearing wall's placement in the building is very important. It increases the structure's rigidity mainly if it is placed somewhere between the edge and the centre bays, by creating a stiff area around it. In the plastic analysis, it is seen how the bearing masonry wall holds the beams attached to it in place, if the seismic action is on its direction. The FB wall however, cannot bear the bending moments it is subjected to at the base. The non-bearing walls show they would be cracked at the corners due to bending moments perpendicular to their plane. The masonry walls have an important role here, but the solution can be enhanced.

References

[1] Costa A A, Penna A and Magenes G 2011 Seismic Performance of Autoclaved Aerated Concrete (AAC) Masonry: From Experimental Testing of the in-Plane Capacity of Walls to Building Response Simulation *Journal of Earthquake Engineering*, 15:1-31, 2011 ISSN: 1363-2469 print/ 1559-808X online doi 10.1080/13632461003642413

[2] Costa A A, Penna A, Magenes G, Galasco A, 2008 Seismic performance assessment of autoclaved aerated concrete (AAC) masonry buildings *The 14th world conference on earthquake engineering. Beijing, China*, 2008 https://scholar.google.ro/

[3] El-Betar S A 2015 Seismic performance of existing R.C. framed buildings *Elsevier Housing and Building National research center (HBRC)* **13**, p 171-180, . http://dx.doi/org/10.1016/j.hbrcj.2015.06.001

[4] Murty C V R and Jain S K 2000 Beneficial influence of masonry infill walls on seismic performance of RC frame buildings *12th World conference on Earthquake Engineering (WCEE) Auckland, New Zeeland.* http://www.iitk.ac.in/nicee/wcee/article/1790.pdf

[5] Teguh M 2016 Experimental evaluation of masonry infill walls of RC frame buildings subjected to cyclic loads *Elsevier Procedia Engineering 171 Sustainable Civil Engineering Structures and Construction Materials*, p 191-200, doi:10.1016/j.proeng.2017.01.326

[6] Li B, Wang Z, Mosalam K and Xie Wenchuan H 2008 Earthquake Field Reconnaissance on Reinforced concrete on reinforced concrete framed buildings with and without masonry infill walls *The 14th world conference on earthquake engineering. Beijing, China*, 2008 https://scholar.google.ro/

[7] Cornado C, Rosell J R, Leiva J and Diaz C 2016 Experimental study of brick masonry walls subjected to eccentric and axial load RILEM International Conference on Materials, Systems and Structures *Civil Engineering Conference segment on Historical Masonry Technical University of Denmark, Lyngby,* p 33–40, 2016. www.rilem.net/publications/proceedings-500218

[8] Marques R and Lorenco P B 2014 Unreinforced and confined masonry buildings in seismic regions: Validation of macro-element models and cost analysis *Engineering Structures Elsevier* vol **64** pp 52-67 https://doi.org/10.1016/j.engstruct.2014.01.014

[9] Yoshimura K, Kikuchi K, Kuroki M, Nokana H, Tae Kim K, Wangdi R, and Oshikata A 2004 Experimental study for developing higher seismic performance of brick masonry walls 13th World Conference on Earthquake Engineering Vancouver, B.C., Canada, paper No. 1597. https://www.iitk.ac.in/nicee/wcee/thirteenth_conf_Canada/

[10] Wijaya W, Kusumastuti D, Suarjana M, Rildova R and Pribadi K 2011 Experimental Study on Wall-Frame Connection of Confined Masonry Wall The Twelfth East Asia –Pacific Conference on Structural Engineering and Construction Elsevier Procedia Engineering **14** p 2094-2102, doi:10.1016/j.proceng.2011.07.263

[11] K. Leng, C. Chintanapakdee, T. Hayashikawa, 2014 Seismic Shear Forces in Shear Walls of a medium Rise Building By Response Spectrum Analysis *Engineering Journal Volume 18 Issue* 4, 2014. http://dx.doi.org/10.4186/ej.2014.18.4.73

[12] CEN EN 1990-2004 Eurocode 0: Basics of structural design, 2004.

[13] CEN EN 1991-1-1-2004 Eurocode 1: Actions on structures - Part 1-1: General actions- Densities, self-weight, imposed loads for buildings, 2004.
[14] CEN EN 1992-1-1-2004 Eurocode 2: Design of concrete structures - Part 1-1: General rules and rules for buildings, 2004.
[15] CEN EN 1996-1-1-2006 Eurocode 6: Design of masonry structures - Part 1-1: General rules for reinforced and unreinforced masonry structures, 2006.
[16] CEN EN 1998-1-2004 Eurocode 8: Design of structures for earthquake resistance. Part 1: General rules, seismic actions and rules for buildings, 2004.
[17] P100-1/2013 Seismic design code–Part 1 - General rules for buildings, 2013 updated P100-1/2019.
[18] CR 6/2013 Masonry design code, 2013.

IOP Conf. Series: Materials Science and Engineering 1283 (2023) 012005 doi:10.1088/1757-899X/1283/1/012005

Technology for utilization and neutralization of municipal waste

V Ezhov[1], V Pakhomov[1], D Fedyanin[1] and I Saikov[1]

[1] Southwest State University, Kursk, Russia

* vlad-pakhomov-03@mail.ru

Abstract: Considered the issues of environmental protection with the simultaneous use for the neutralization and disposal of urban waste containing substances of organic origin, due to solar energy and atmospheric precipitation. At present, in the Russian Federation, waste makes up 90–98% of all extracted natural resources, that is, the industry works mainly for the production of waste. Already 8 billion m^3 of industrial and municipal waste has accumulated. Landfills are 50–90% full on average across the country. The volume of waste from livestock enterprises and poultry farms in the form of manure, droppings and sewage is about 700 million m^3 per year, and more than 2 million hectares of land are occupied for storing this waste. Therefore, it is planned until 2020–2025 to introduce a complete ban on the storage of municipal solid waste at more than 200 landfills and dumps, where they have accumulated 31 billion tons

1. Introduction

At present, in the Russian Federation, waste makes up 90–98% of all extracted natural resources, that is, the industry works mainly for the production of waste. Already 8 billion m^3 of industrial and municipal waste has accumulated. Landfills are 50–90% full on average across the country. The volume of waste from livestock enterprises and poultry farms in the form of manure, droppings and sewage is about 700 million m^3 per year, and more than 2 million hectares of land are occupied for storing this waste. Therefore, it is planned until 2020–2025 to introduce a complete ban on the storage of municipal solid waste at more than 200 landfills and dumps, where they have accumulated 31 billion tons [1-3].

The tense situation that has developed in recent years with the environmental situation in the areas where landfills and unauthorized dumps are located has increased the attention of society and the state to the need to take quick and decisive measures to solve this problem, which consists in the disposal and disposal of solid municipal waste (MSW).

2. Experimental procedure

At present, the most common methods for solving this problem are thermal technologies for the neutralization and processing of MSW. Thermal methods of MSW disposal are one of the bulk processing methods. The advantage of the method is a significant reduction in the volume of the processed material. When waste is incinerated, their weight is reduced by 85–90%. Ash and slag formed during combustion are removed to landfills. The gases formed during thermal processing are purified and removed to the atmosphere. There are several options for thermal methods: combustion, gasification and pyrolysis [2].

Combustion is carried out in an oxidizing environment at a temperature not lower than 600°C. In this case, the process proceeds auto thermally and does not require the introduction of additional fuel to maintain combustion. The method is most often used in the operation of incinerators. The advantages of the method are the simplicity of the organization of the sludge management, the compactness of the equipment, and the low cost of cleaning the exhaust gases. The disadvantage of this method is the need for preliminary sorting of waste. They should not contain compounds of phosphorus, halogens and sulfur. Otherwise, the combustion process will generate highly toxic carcinogenic gas emissions containing dioxins and furans. Incineration is used in the processing of MSW at waste incineration plants. In Russia, some of the incineration plants purchased abroad do not work, since they are not adapted to the combined municipal waste that prevails in our country.

Gasification is used for the disposal of solid and pasty carbon-containing waste. It consists in the processing of waste at a temperature of (600-1100)°C with water vapor, oxygen or carbon dioxide. A reducing atmosphere is maintained in the reactor, which excludes the formation of sulfur or nitrogen oxides. Gasification produces partial oxidation products such as CO, CO_2, H_2O, HCl, aldehydes, phenols, hydrocarbons. This process can be carried out if the amount of released toxic components does not exceed the maximum allowable concentration (MPC norm). Generator gas can be used locally as a low-calorie fuel. Pyrolysis is a high temperature process. It is carried out at a temperature of (700–1050)°C in an oxidizing environment or without access to air.

The literature contains developments on the use of higher temperatures and the use of plasma technologies up to 1700° C. Pyrolysis is based on deep oxidation of organic residues. When carrying out pyrolysis, the processes of drying, dry distillation, and gasification are interconnected. CO_2, H_2O, CH_4 are obtained as final products.

The main disadvantages of the known methods of gasification and pyrolysis of solid waste are:

- the impossibility of using natural factors (energy of the sun and atmospheric precipitation) for the disposal and neutralization of solid waste;
- the need to burn fuel from an external source in an underground mobile combustion chamber;
- the presence of emissions of a significant part of gaseous products obtained as a result of reactions occurring in the waste array through the outer surface of the waste array into the surrounding atmosphere;
- the need to use complex and expensive equipment for their implementation, which significantly reduces their environmental and economic efficiency.

For the utilization and neutralization of organic components in the massifs of municipal waste, the authors have developed a basic technological scheme (technological line) for the implementation of the proposed technology for the neutralization and disposal of solid waste, shown in figure 1 and 2 (figure 1 - general view, figure 2 - main nodes) [3].

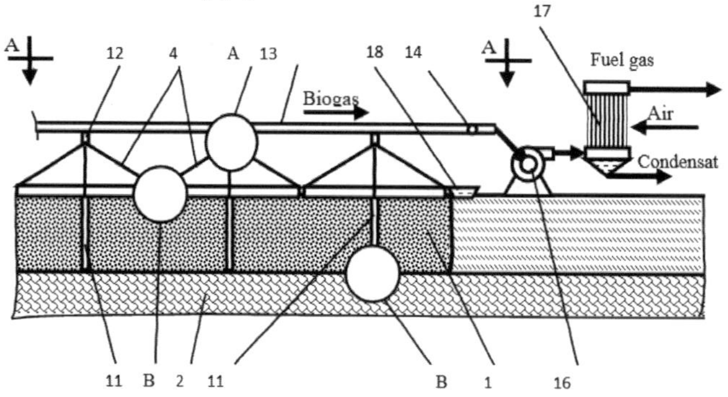

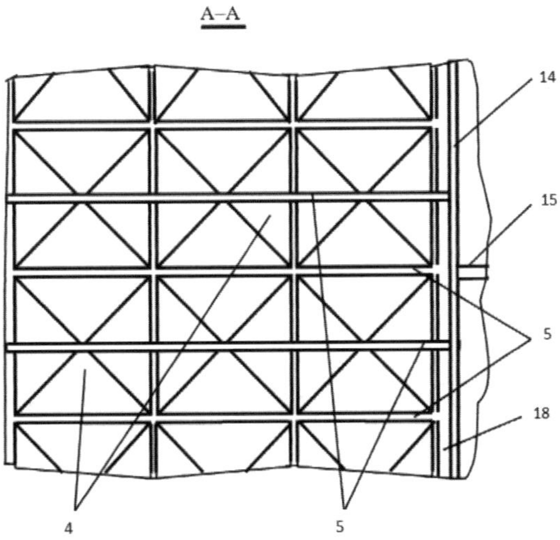

Figure 1. Basic technological scheme for the implementation of technologies for neutralization and utilization of solid waste in the section A-A (general view): 1 - a section of an array of household waste; 2 - the bottom of the 2nd landfill; 4 – N transparent sealed pyramidal caps; 5 – grooves; 11 – outlet vertical perforated pipes; 12 - inlet pipe; 13-row gas manifold; 14 – common gas collector; 15 – suction gas pipeline; 16 – compressor; 17 – tubular air cooler; 18-distribution tray.

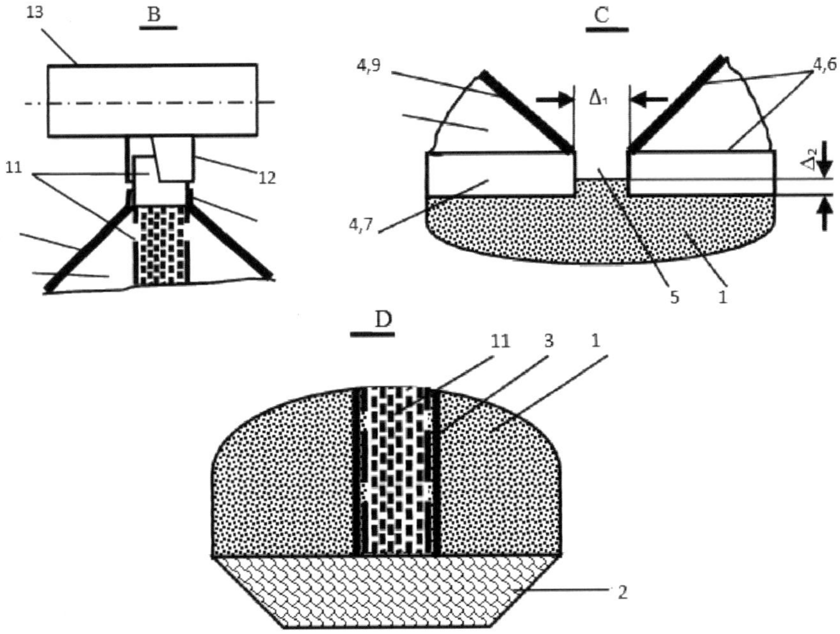

Figure 2. Basic technological scheme for the implementation of technologies for neutralization and utilization of solid waste in the context of B, C, D (main units): 1 – section of the massif of household waste; 2 – bottom of the 2nd landfill; 3 – N wells 3; 4 – N transparent sealed pyramidal caps; 6 – frame of the cap; 7 - square frame; 8 – top ring of the cap; 9 – inclined ribs; 10 – transparent shell; 11 – outlet vertical perforated pipes; 12 - inlet pipe; 13-row gas manifold.

The proposed technology for neutralization and utilization of the solid waste array is carried out in the sequence described below.

Preliminarily, stationary equipment is installed near the landfill for the technological line for the neutralization and utilization of MSW. It is advisable to choose the location of the technological line in the immediate vicinity of the water supply source. The stationary equipment of the processing line includes a compressor, an air tube cooler, a distribution tray, a gas collector, a storage tank and a feed pump.

Further (figure 1), in the area of the massif 1 of the landfill, wells 3 are drilled, where vertical perforated pipes 11 are inserted, reaching the bottom end of the base 2 of the landfill, after which the free space of the wells 3 is filled up.

Then, in the above-mentioned section of the array 1, N pre-assembled, transparent sealed pyramidal caps 4 are installed with gaps between themselves horizontally with a width of $\Delta 1$ and a depth of immersion in the array 1 - $\Delta 2$, forming grooves 5. In this case, the immersion depth $\Delta 2$ must ensure sufficient tightness caps 4, and the width of the gaps $\Delta 1$ and their depth should ensure reliable moistening of the area and thickness of the array1 under the caps 4.

After that, on the upper ends of the pipes 11, the intake pipes 12 of the ordinary gas manifold 13 are put on, they are connected to the common gas manifold 14, which is connected through the suction gas pipeline 15 to the compressor 16.

The area of the caps 4 is found on the basis of the optimal distance of the suction of biogas by the pipe 11 and the moistening of the massif 1 from the grooves 5, and the angle of inclination of the ribs 9 should be not less than the angle of repose of the water. The number of pyramidal transparent caps 4 N and the corresponding number of wells 3 with outlet pipes 11 is limited by the aerodynamic resistance of the suction part of the gas pipelines of the installation.

The neutralization of organic components of the landfill is carried out in the warm season directly in the massif 1 of the stored waste using solar heating and rain moistening (in case of insufficient natural moisture, water is supplied to grooves 5 from an external source of water supply through a feed pump and a distribution tray 18) to obtain biogas (methane) and its subsequent output through the outlet pipes 11 and the pores of the array 1 into the space under the bell 4 and gas pipelines 3.

When the massif 1 is saturated with rainwater or water from a water supply source and it is heated by the sun's rays through the transparent shell of the pyramidal caps 4, in the thickness of the massif 1, something similar to a wet sediment is formed in the "methan tank" (device for anaerobic fermentation of liquid organic waste with the production of methane) with a temperature of (20-30)°C, which is quite close to the optimal temperature of anaerobic fermentation (30-50)°C. The temperature in the thickness of the massif 1 can also increase due to exothermic reactions occurring between its components.

At the same time, along with humidification, the massif 1 is saturated with impurities present in rain or make-up water (CO_2, NO_X, SO_X, salts of Ca, Mg, etc.), as a result of the processes of absorption, adsorption and chemisorption that occur with the components of the massif of household waste (protein particles, fats, sand, clay, etc.), and in the massif 1 during the decomposition of fats, proteins, mineral salts, etc., CO_2 is formed, and rainwater decreases the pH of the massif 1, which also intensifies the processes of CH_4 formation. As a result of the interaction of the above factors, the organic components of the raw massif 1 are neutralized by anaerobic digestion, which is the main method for neutralizing raw sewage sludge, which has approximately the same composition as the domestic waste array [4-6]. At the same time, as a result of the decay of organic substances from household waste and the interaction of decay products with carbon dioxide, methane is obtained as the main products [6-9].

Methane is formed as a result of the reduction of CO2 or the methyl group of acetic acid:

$$4AH_2 + CO_2 \xrightarrow{ferments} 4A + CH_4 + 2H_2O \quad (1)$$

where AH_2 is an organic substance serving as a hydrogen donor for methane-forming bacteria (fatty acids other than acetic and alcohols other than methyl).

In addition, many types of methane-forming bacteria oxidize molecular hydrogen, which is formed in the acid phase by the reaction:

$$4H_2 + CO_2 \xrightarrow{ferments} CH_4 + 2H_2O + energy \tag{2}$$

Microorganisms using acetic acid and methyl alcohol carry out the reactions:

$$CH_3COOH \xrightarrow{ferments} CH_4 + CO_2 + energy \tag{3}$$

$$2CH_3OH \xrightarrow{ferments} CH_4 + CO_2 + H_2 + energy \tag{4}$$

In this case, if there is free CO_2 in the array 1, the rate of reactions (1), (2) increases, and the rate of reactions (3), (4) decreases, which increases the proportion of methane in the produced gas.

The resulting biogas is fed by the compressor 16 into the tubes of the air cooler 17, which is cooled by the outside air due to natural draft, where it is cooled with the condensation of a significant part of water vapor and heavy hydrocarbons.

The purified and cooled biogas, consisting mainly of CH_4, enters the gas collector, and the condensate is directed to the storage tank. In this case, as a result of the operation of the compressor 16, a certain rarefaction is created in the cavity of the caps 4, which increases the rate of detachment of the molecules of the biogas components from the surface of the array 1 under the cap 4. And this leads to an increase in the speed of movement of these molecules in the pores of the array 1 and the difficulty of their movement towards the surface of the grooves 5, which prevents them from entering the atmosphere. In addition, the constant presence of moisture on the surface of the grooves 1 also prevents molecules of biogas components from entering the atmosphere.

The process of neutralization of a section of the array 1 is carried out during once of the warm period of the year. The duration of the process depends on the average temperature of the warm period, the thickness and porosity of the massif 1, the content and nature of the organic components of the waste.

At the end of the disposal of MSW, dismantle all equipment simultaneously with the removal of fermented household waste. Then the technological line is transferred to the next section of the waste disposal site.

To speed up the landfill deactivation process, it is possible to arrange several simultaneously operating technological lines in the areas of array 1.

3. Conclusions

The proposed technology, due to the use of natural factors (solar heat and atmospheric precipitation), along with improving the ecological situation in the places of disposal of household waste, ensure the utilization of their most dangerous (organic) part with the production of biogas (fuel gas – methane).

The technology is implemented without significant energy costs (energy is spent only on the drive of the compressor and feed pump).

The produced methane can be used as fuel for heat generators.

Condensate, depending on the type and concentration of hydrocarbons in it, is sent for further processing or discharged with feed water to moisten the massif.

The fermented massif after separation of foreign objects from it is used as a highly effective fertilizer for agriculture.

In comparison with the well-known thermal methods of neutralization and disposal of MSW, the proposed technology, due to the above factors, significantly increases the reliability, environmental and economic efficiency of the process of neutralization and disposal of an array of household waste.

References

[1] Khoroshavin L, Belyakov V, Svalov E 2016 Basic technologies of industrial and solid municipal waste processing. Tutorial. - Ekaterinburg: URFU

[2] Semenova N 2009 Industrial ecology - Moscow, Publishing center "Academy", 528
[3] Ezhov V 2017 Method and devices for disposal and disposal of household waste. Patent of Russia No. 2630456. Bul. No. 25
[4] Nenizescu K 1968 General chemistry. - Moscow: Mir, 816
[5] Yakovlev S 1976 Moscow: Gosstroyizdat, 453
[6] Ezhov V S, Semicheva N E, Pakhomova E G, Polivanova T V 2021 Ecology and Industry of Russia **25(2)**, 4–7 doi: 10.18412/1816-0395-2021-2-4-7
[7] Ezhov V S, Semicheva N E, Bredikhina N, Semerinov V G, Ezhova T 2019 Chemical and petroleum engineering **55 (5-6)**, 514-521 doi: 10.1007/s10556-019-00654-1
[8] Polivanova T V, Merdeev R R 2013 Proceedings of the Southwestern State University. Series: Engineering and Technology **4**, 136-139
[9] Dolzhikov P N, Psyuk V V, Shubin A A 2012 Plastic materials and structures in construction. Study guide. Alchevsk, DonGTU Publishing House, 188

Studies on heavy duty pavements reinforced with steel fibers for port infrastructures

P Ionițoaiei[1,2*] and A Rotaru[1]

[1] Gheorghe Asachi" Technical University of Iași, Romania
[2] PORR Construct, București, România

* ionitoaiei.pavel@gmail.com

Abstract. Steel fiber reinforced concrete (SFRC) for industrial floors has become a state-of-the-art construction method in all industries worldwide, including the construction of port platforms. The first patent that refers to a dispersed reinforced concrete element with steel fibers there exists since 1874 and was patented in the USA (California) by A. Bernard who proved the improvement of the strength of the concrete by adding some uneven steel scraps. Steel fibers have been created to improve the durability and flexibility of concrete under high loads, over long periods of time, providing three-dimensional reinforcement of concrete. Steel fibers can be used as the sole form of reinforcement or in combination with reinforcement bars or welded wire mesh.

1. Introduction

As a short history of steel fiber reinforced concrete, the first patent that refers to a dispersed reinforced concrete element with steel fibers there exists since 1874 and was patented in the USA (California) by A. Bernard who proved the improvement of the strength of the concrete by adding some uneven steel scraps. In 1927, also in California, G. C. Martin patented the construction of reinforced concrete pipes with steel fibers. Meischke – Smith in 1920 and Etheridge in 1933 highlighted the correlation between fiber shape and enlargement of adhesion. The first used flat wires and twisted wires with flat faces, while the second used annular fibers, with different sizes and diameters, to improve the resistance to cracking of the concrete.

Steel fibers have been created to improve the durability and flexibility of concrete under high loads, over long periods of time, providing three-dimensional reinforcement of concrete. Fiber-dispersed reinforced concrete cannot completely replace ordinary reinforced concrete.

There are, however, areas of use in which reinforced concrete with fibers can be used alternatively or in addition to the classic reinforced one, offering constructive and economic advantages, as in the case of Constanța Southport Development.

The paper presents data on the extension and modernization of the infrastructure for the transport of Ro-Ro goods, general cargo and, in the long term, for containers. The investment is in line with the strategic objectives in the transport sector at European and national level.

The expansion of the terminal in the development area A, MOL II-S Port Constanța South can be achieved by extending the multifunctional operational and storage platforms, the scanner platform for containers, transformer stations, the canyon and the arrangement of platforms behind berth 128 for RO-PAX activities and Project Cargo type goods. Main works to be executed:

- 63.000 m^3 of BcR 5.0 concrete platform with a load-bearing capacity of $20t/m^2$ for containers and portal cranes,

- 60 km of utilities, electrical and water network for the operational area and ships supply,
- Hydrotechnical works for new ship-to-shore cranes.

The initial design solution had a concrete pavement reinforced with over 7,700 to of cut and bent steel rebars which implies a large effort of manpower and duration of execution as manual rebar works and pouring of the concrete slabs and a high CO_2 emissions impact due to big amount of steel to be produced and transported to the site.

The initial design solution had the following structure:

- 45 cm of BcR 5.0 concrete pavemet reinforced with 120kg rebar/m^3
- 30 cm of stabilized ballast layer (base layer)
- 40 cm of crushed stone layer 0-63 (foundation layer)
- geotextile and geogrid layer
- compacted foundation soil

To increase productivity and reduce the large labour effort and CO_2 emissions, but also to ensure the stability and functionality requirements of port platforms, a new optimized solution had to be developed. Steel fiber reinforced concrete offers an alternative to conventional reinforcement, having as advantage the time and low costs of execution of works.

By replacing the cut and bent steel from the pavement with steel fibers [1-5] an optimized solution was developed, which allowed a mechanical casting [6] of the concrete pavement slabs, increasing on-site productivity, reducing the execution time and also significantly reducing the amount of CO_2 emissions.

The paper presents the case study of the project Constanța Southport Development, from the design phase, testing and execution phase of the optimized solution on site.

Figure 1. Site overview at the beginning of the works.

2. Materials and methods

The behavior of ground slab is different than that of a roof slab as it rests directly on the soil. This is why, if at all conventional reinforcement is added to ground slab, it is added only for crack control at the top layer, preferably in the top one-third layer of the slab.

Reinforcement below the mid depth is uncalled for both behaviorally and economically. This leads to very limited crack control in conventional ground slabs, as the top reinforcement is localized and there is no mechanism for arresting cracks below the mid depth.

Also, the moment resistance is provided completely by the concrete in the slab. Steel fibers, are spread throughout the slab thickness, and hence can arrest the cracks at the location they originate. Also, the fibers provide a post-cracking moment resisting capacity, increased energy absorption and impact resistance. The extent to which these benefits are derived from SFRC depends on the type of fiber, and the fiber content (fiber percentage by volume and number of fibers).

The length to diameter ratio mentioned earlier is an important governing factor. As the stress transfer from the concrete matrix to the fibers happens by interfacial shear, the greater the interfacial surface area, greater is the fiber efficiency. Hence fibers with greater aspect ratio have greater efficiency.

However, there is a limit to this, as very high aspect ratio will reduce the workability of the concrete mix. Most practical applications use an aspect ratio less than 100. The stress in SFRC is shared by the concrete matrix and fibres until the concrete cracks. After cracking, the total stress is gradually transferred to the fibres.

The resistance to pull-out of fiber will depend, apart from the aspect ratio, on the shape of fiber. Non round fibres, fibres with hooked ends provides more pull-out resistance compared to plain round fibres. A pull-out type of failure gives gradual and ductile failure. To prevent this, we chose fibres of triple hooked ends. It was used fibers of 0.90 mm thickness and 60 mm length, thus having an aspect ratio of 65.

2.1. Technical performance of steel fibers in concrete pavements

- Bending force [7, 8] – During the critical post-cracking behavior of the concrete matrix, the superiority of the steel fibers can be demonstrated, the metal fibers with beaks requiring an increased energy intake for extraction from the matrix. For a fiber content that varies from 20 kg/m^3 to 45 kg/m^3 the growth order of the result at residual bending is approximately 60%.
- Durability [7, 8] – Steel fibers have been used in various applications involving a wide range of environmental conditions. The metal fibers performed well because they are short and discontinuous and improve the corrosion resistance caused by the effects of oxidation.
- The resistance to fatigue by bending [7, 8] concrete with steel fibers is higher than that of simple concrete. Steel fibers prevent the development of major cracks in concrete subjected to cyclical loads. Improved fatigue resistance allows steel fibers to be used efficiently in pavements.
- Impact resistance [7, 8] – Steel fibers increase the impact resistance of concrete through its inherent ability to increase the strength of the concrete matrix and absorb energy.
- Abrasion resistance [7, 8] – Steel fibers are randomly oriented and dispersed evenly in the concrete mass. The result is that the concrete surface is better reinforced so that the abrasion resistance of concrete is increased.

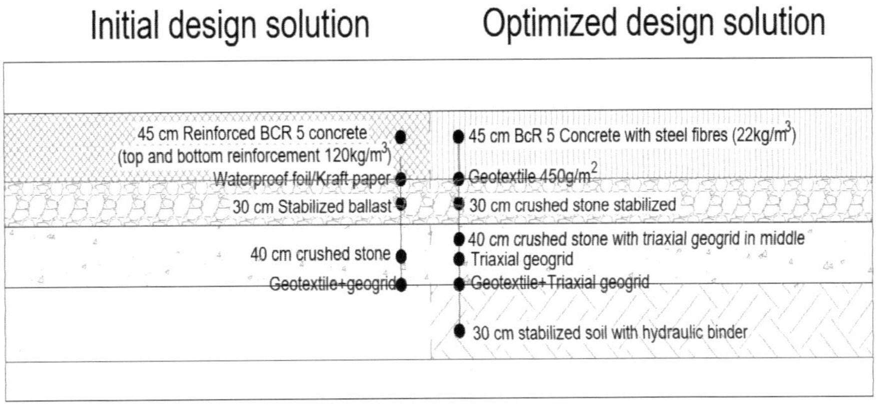

Figure 2. Initial / Optimized design solution.

The optimized design solution compared to the initial design solution (Figure 2):

- 45 cm of BcR 5.0 concrete pavemet reinforced with 22kg SFRC/m^3;

- 30 cm of stabilized crushed stone 0-31.5 (base layer);
- 40 cm of crushed stone layer 0-63 (foundation layer);
- geotextile and triaxial geogrid layer;
- 30 cm in situ stabilized foundation soil.

2.2. Logistics, materials and equipment for the preparation of concrete reinforced with steel fibers (BcR 5.0 – road concrete)

In order to produce the concrete reinforced with steel fibers, it was necessary, besides the concrete station itself, an automatic fiber dispenser to dose the quantity established by the recipe directly on the conveyor belt so that the fibers reach the mixer at the same time as the aggregates, thus achieving a homogeneous dispersion of the steel fibers in the concrete matrix (Figure 3).

A sufficient number of tests were performed on a trial section in situ, the characteristics of the fresh concrete were determined and the optimal times for the production of concrete reinforced with steel fibers were established.

Figure 3. Concrete plant with automatic steel fibers dozer.

The following BcR 5.0 road concrete pavement recipe was established:

- Cement I52,5R 370 [kg]
- Water 164 [L]
- Sika Plasticizer 640 BV - 0.6% 2.22 [L]
- Sika LPS A 94 - 0.25% 0.93 [L]
- Steel fibers kg/m3 22 [kg]
- 0/4 river aggregates 560 [kg]
- 4/8 quarry aggregates 387 [kg]
- 8/16 quarry aggregates 389 [kg]
- 16/31.5 quarry aggregates 544 [kg]
- Density [kg/m3] 2439 [kg]

Characteristics of steel fibers used for the recipe:

- Tensile strength 2,300 N/mm^2
- Young's modulus 200,000 N/mm^2
- Strain at ultimate strength 6%
- Length 60mm / Diameter 0.90mm
- Aspect ratio (l/d) 65
- Network 3124 fibers/kg, 4400 m/m^3 at 22 kg/m^3

For a better dispersion of steel fibers in the concrete matrix but also for the way of behavior and tensile strength, glued steel fibers were used for this type of concrete pavement, bundled with water-soluble glue (Figure 4). The glue helps avoiding fibers balling during mixing and ensures a homogeneous distribution of fibers throughout the concrete mix.

Figure 4. Steel fibers dispersion on fresh concrete matrix.

3. Results

Samples for testing were taken and tested at 7 and 28 days, both for compression and bending strength. The results are shown below in Figure 5 and Tables 1 and 2. The design strengths, after 28 days, for this type of road concrete *BcR 5.0* according with initial recipe is minimum 50 N/mm^2 for compression strength (Rc) and minimum 5.50 N/mm^2 for bending strength (Rinc), shown in Table 3.

Table 1. Compression strength results for steel fiber reinforced concrete.

Days	Sample ID	Compression strength (Rc)[N/mm^2]	Average	% Design strength
7days	2 BCR _1	48.99 49.55	49.27	98.54%
28days	2 BCR _2	59.86 58.74	59.3	118.60%

Table 2. Bending strength results for steel fiber reinforced concrete.

Days	Sample ID	Bending strength (Rinc) [N/mm^2]	Average	% Design strength
7days	2 BCR _1	5.69 5.57	5.63	102.36%
28days	2 BCR _2	6.98 7.19	7.08	128.73%

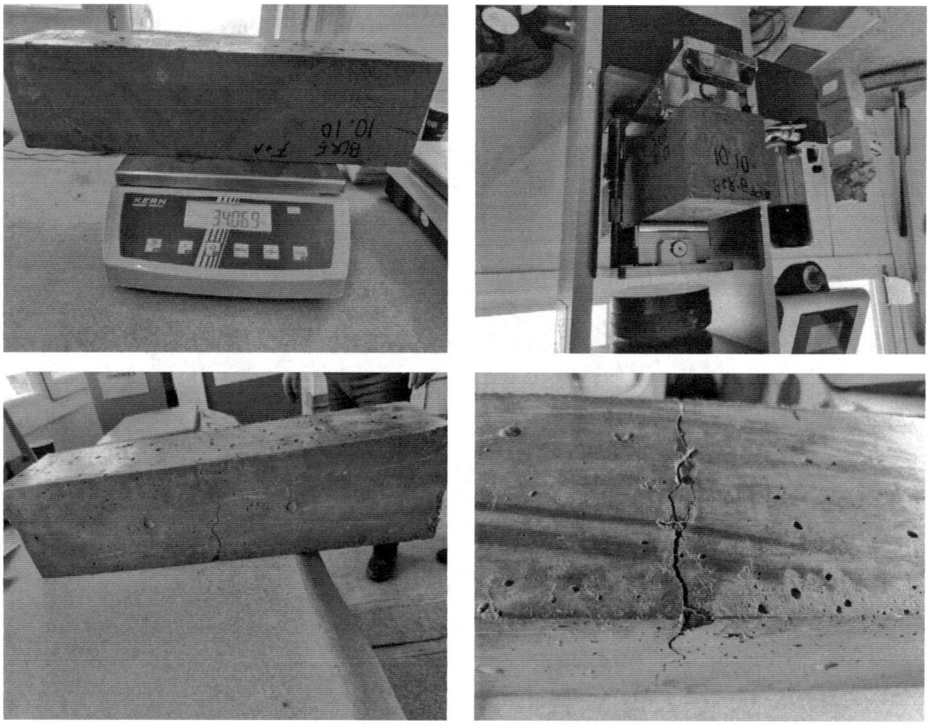

Figure 5. Tested samples.

Table 3. Design strengths for BcR 5.0 without reinforcement.

Days	Sample ID	Compression strength (Rc)[N/mm^2]	Bending strength (Rinc) [N/mm^2]
7days	1 BCR _0	42.27	5.10
28days	1 BCR _0	52.02	5.65

Performance of steel fibers in concrete pavements applied to the project:

- Fast preparation and laying of the concrete layer (Figure 6),
- Substitution of conventional cut and bent reinforcement,
- Reducing the labor cost that is normally associated with fixing the metal,
- Easier pouring of concrete (Figure 7),

- Increased reliability of joints during the life of the pavement,
- Reductions in the cost of maintenance,
- CO_2 emission of steel fiber production is lower than traditional steel rebar process (Table 4).

Table 4. CO_2 emission table for traditional steel rebar [9] versus steel fibers (2021 indicators).

CO_2 emission	kg/m^3	CO_2/kg	CO_2/m^3	BcR 5.0 m^3	CO_2 emissions [to]
Steel rebar	120.00	1.91	229.20	63,000.00	14,439.60
Steel fibers	22.00	0.96	21.12	63,000.00	1,330.56

Figure 6. BCR5 reinforced with steel fibres poured on site.

Figure 7. Site overview during the puring of BCR5 reinforced with steel fibres.

4. Conclusions

Due to its characteristics: flexural resistance, fatigue resistance, resistance to damage in case of impact, high ability to absorb and dissipate energy, high permeability, abrasion and exfoliation resistance, elimination of cracks, steel fiber reinforced concrete is an ideal material for use in port pavements.

Steel fiber reinforced concrete offers an alternative to conventional reinforcement, having as advantage the time and low costs of execution of works.

Steel fibers improve the durability and flexibility of concrete under high loads, over long periods of time, improve ductility, energy absorption properties and impact resistance, ensuring three-dimensional reinforcement of concrete.

In practice, it is noticed the increase of the bending resistance of the reinforced concrete element with steel fibers, due to the fact that the distribution of efforts in the bent area is almost constant.

The development and implementation of the solution for dispersed reinforcement with steel fibers of the pavement has brought many advantages in carrying out the execution of the works. Productivity increased from an average of 250÷300 m^3 of concrete estimated to be poured manually to an average of 900 m^3 per day mechanized cast, also increasing the quality of the works performed.

A major aspect of using steel fibers instead of traditional reinforcement is the impact on the environment by significantly reducing CO_2 emissions. As shown in Table 4, by using steel fibers in this project, the emission of CO_2 was reduced by more than 13,000 tons, which emphasizes that steel fibers are much more environmentally friendly than traditional reinforcement.

Regarding the mechanical characteristics of the dispersed reinforced concrete with steel fibers it can be seen from Tables 1, 2 and 3 that the reinforced concrete with steel fibers reaches the design resistances even after the first 7 days, creating the advantage of working the joined strips even before the passage of the 28 days according to the norms.

References

[1] Knapton J 2008 The Structural Design of Heavy Duty Pavements for Ports and Other Industries, Edition 4, *Interpave*

[2] SR EN 14889-1:2007, Fibers for concrete. Part 1: Steel fibers. Definitions, specifications and conformity

[3] NE 012/2-2010 Regulations for the production of concrete and the execution of concrete works, reinforced concrete and prestressed concrete, Part 2: Execution of concrete works

[4] GP-075-02 Guide for establishing performance criteria and compositions for reinforced concrete dispersed with metallic fibers

[5] Murugesan M and Dashrath Rajpurohit, 2015, Advantages of steel fibre reinforced concrete in industrial floors, *Int J Res in Eng & Techn (IJRET)*, **4** (10).

[6] NE 014-2002 Regulations for the execution of roads made of cement concrete in the systems fixed and sliding formworks

[7] Decision: 99/469/EC, Product family, product / intended use: Products related to concrete, mortar and cement paste (1/2): – Fibers for structural uses in concrete, mortar and cement paste

[8] Cioc A and Badiu M, 2011 Steel fibers for concrete reinforcement non-compliance with the harmonized European standard SR EN 14889/1:2007, The 17th SINUC 2011 National Construction Machinery Symposium. UTC Bucharest

[9] Worldsteel Association: Sustainability Indicators for 2021 https://worldsteel.org/steel-topics/sustainability/sustainability-indicators (accessed on 02.02.2023)

Eco-concrete for obtaining "green" construction elements

B I Luca[1], A Panțiru[1], A Timu[1], M Bărbuță[1], L I Diaconu[1], M Rujanu[1] and A C Diaconu[2]

[1] "Gheorghe Asachi" Technical University of Iasi, Faculty of Civil Engineering and Building Services, Iasi, Romania
[2] Romanian Association of Earthquake Engineering, Romania

* livia-ingrid.diaconu@academic.tuiasi.ro

Abstract. The building materials branch of the construction industry is continuously developing due to occurrence of new products which seek to improve the "green" character of buildings by using different types of wastes. Prefabricated products of different sizes such as blocks, panels, beams and other spatial elements can now be obtained with eco-materials. Concrete is the most used material in the building industry and thus it is necessary to diminish its environmental pollution. There are different ways for improving the "green" character of concrete, such as the addition of waste products of different types (fly ash, silica fume, slag, shredded tyre rubber, agricultural powder wastes, etc.) or by replacing some of the components (i.e. cement, aggregates) with eco-materials. Materials with pozzolanic properties (fly ash, silica fume, banana leaves powder, volcanic tuff, etc.) can, in certain proportions, replace the cement. In order to replace the aggregates, the substitution materials are obtained with sizes which are corresponding to the aggregate sort (polystyrene granules, shredded plastic recipients, agricultural wastes, etc). In the current article, the experimental results obtained on types of eco-concrete used for producing reinforced concrete beams are presented. The eco - concrete used for pouring the reinforced concrete beams contains the following materials: fly ash as replacement for the cement, metalic fibers, polyester fibers and hemp fibers. The beams models were tested in two point bending. The strength capacity and type of failure were analysed and compared as a function of the type of eco-concrete.

1. Introduction

The innovative elements refer to the carrying out of several tests of concrete recipes with different additions in accordance with the subsequent possibilities that the same work protocol can be applied on a large scale by companies specialized in the field of concrete production and construction materials [1-3]. The most common modern cement mixture, known as Portland cement has been used for near 200 years, but it is not very friendly to the environment as cement production led in 2021 to 7% of emissions of carbon dioxide, worldwide.

The main objective of the experimental research in this paper is the analysis of the physical-mechanical properties of some types of concrete that want to be ecological, using fly ash as a substitute for cement and different fibers [4].

The concrete with waste used for obtain reinforced concrete beams presented the characteristics closed to that of a regular concrete of class C20/25 and was presented in [5].

This study focuses on improving and add to the current level of knowledge in the field of construction materials, especially concrete by analyzing the behavior of some concretes and concrete elements made with ecological recipes by using renewable natural additives with the aim of reducing the production cost and obtaining a material more effective from the point of view of the physical-mechanical characteristics than those of the classic concrete studied up to now.

The topic of this paper is well documented in the specialty literature, the published papers have explored a wide range of topics related to eco-concrete, including material properties, its impact on the environment, and its potential applications in the construction industry. Some papers have focused on the use of recycled materials and their effects on the strength and durability of the materials, other papers have looked at the thermal performance of eco-concrete and its potential to reduce energy consumption in buildings [6-8].

Eco-concrete is a sustainable alternative to traditional concrete that aims to reduce the environmental impact of construction. It is made by incorporating recycled materials, such as fly ash, metalic fibers, polyester fibers and hemp fibers, into he concrete mixture, which reduce the amount of cement required and results in lower carbon emissions. Eco-concrete also has the potential to improve the thermal performance of buildings and reduce energy consumption. From a sustainability perspective, eco-concrete has the potential to reduce the carbon footprint of construction, conserve natural resources, and contribute to a more sustainable build enviroment. As the construction industry seeks to reduce its environmental impact, eco-concrete is emerging as an important tool for echiving these goals.

2. Experimental program

2.1. Mixes composition

For the preparation of the eco-concrete, a witness mix (CCI) was used, containing the following materials:

- cement type CEM II 42.5 - 360 kg/m^3;
- river aggregates: sand sort 0 – 4 mm – 803.16 kg/m^3;
- sand sort 4 - 8 mm – 384.12 kg/m^3;
- sand sort 8 - 16 mm – 558.72 kg/m^3;
- water 180 l;
- Master Glenium SKY 617 superplasticizer / water reducing additive in a dosage of 1% from cement's amount

In the experimental program, a series of cement concrete mixes were prepared, where the cement was replaced with 10% (BC10) till 40% (BC40) with fly ash (FA) from CET Holboca Iassy. The receipts are presented in table 1.

Table 1. Witnesses' receipts BC and BC10-BC40 with thermo power plant ash.

Receipt	Cement [kg/m^3]	Water [l/m^3]	Thermo power plant ash [kg/m^3]	Sand [kg/m^3]	Aggregate [kg/m^3]	Aggregate [kg/m^3]	Additive [kg/m^3]
BC	360	180	0	803	384	558	3.6
BC10	324	180	36	803	384	558	3.24
BC15	306	180	54	803	384	558	3.06
BC20	288	180	72	803	384	558	2.88
BC30	252	180	108	803	384	558	2.52
BC40	216	180	144	803	384	558	2.16

2.2. Experimental set up

During the research, several types of recycled materials were studied in different dosages to determine the physical and mechanical characteristics of the concrete obtained with each of the materials used for each dosage separately, in order to compare the results, thus aiming to create an ecological concrete, with costs of lower production than ordinary concrete, with a less negative impact on the environment, but with strengths comparable to those of ordinary concrete, then trying to use the concrete obtained in structural elements [9,10].

The concrete compositions studied in order to determine the compressive strength; fc was:

- Concrete with fly ash and hemp fibers **BFC.** The concrete compositions were prepared as follows:
  - fly ash replaced 10% to 40% of the cement dosage;
  - hemp fibers used in proportion of 0.25% with a length of 30 mm.
- Concrete with fly ash and PES fibers **BPES30.** The concrete compositions were prepared as follows:
  - fly ash replaced 10% to 40% of the cement dosage;
  - polyester fibers used in proportion of 0.25% with a length of 30 mm.
- Concrete with fly ash and metallic fibers **BFAFM.** The concrete compositions were prepared as follows:
  - fly ash replaced 10% to 40% of the cement dosage;
  - metallic fibers used in proportion of 1% with a length of 50 mm.

After 28 days the compressive strength was determined according to [11] on cubes of 150 mm sizes. The results are presented in table 2.

Table 2. Compressive strengths of mixes with fibers.

Mix	Dosage	fc N/mm^2
Concrete with hemp fibers	10%	28.36
	15%	27.12
	20%	23.70
	30%	21.10
	40%	17.42
Concrete with Pes fibers	10%	29.8
	15%	24.14
	20%	27.16
	30%	20.93
	40%	18.24
Concrete with FM fibers	10%	30.72
	15%	27.05
	20%	30.56
	30%	30.12
	40%	22.08

For the reduced scale reinforced concrete beams obtained with each of the recycled materials, the following compositions were used:

- **BFC**- Concrete with 10% fly ash as a substitute for cement and hemp fibers in a proportion of 0.25%, with a length of 30 mm;
- **B PES30** - Concrete with 10% fly ash as a substitute for cement and polyester fibers (PES) in a proportion of 0.25%, with a length of 30 mm;
- **BFAFM** - Concrete with 10% fly ash as a substitute for cement and metallic fibers in proportion of 1%, with a length of 50 mm.

2.2.1. Framework, reinforcement and casting of reduced scale reinforced concrete beams. For the formwork, reinforcement and casting of concrete elements with recycled materials, the provisions of the norms in force were taken into account. The shape and reinforcement model of the beams made of the concretes with additions of ecological waste are shown (Figure. 1) below. The samples to be tested are made in patterns of specified shape, dimensions and tolerances [12]. After the re-homogenization of the fresh concrete sample, the prepared frameworks are filled, the concrete is compacted, the surface leveled and their visible marking is completed (figure. 2). The number of layers of concrete applied to fill the patterns, as well as the compaction method, is chosen depending on the consistency of the concrete.

The preparation and preservation of the samples for resistance tests was done according to the standard: SR EN 12390 - part 2. [13].

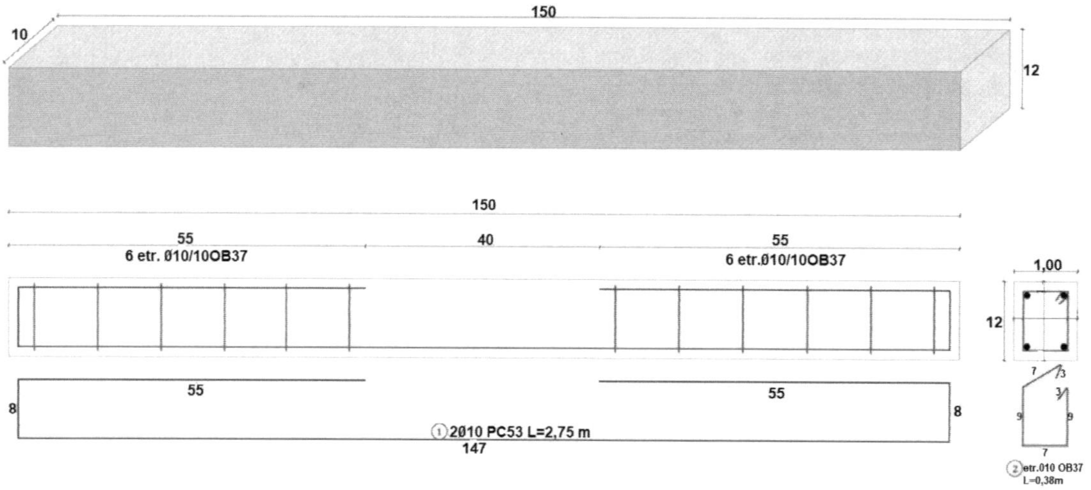

Figure 1. Shape and reinforcement of concrete beams with additions of recycled materials.

Figure 2. Mounting the reinforcement in the formwork (left), freshly poured concrete beam (right).

After the preparation, casting and storage of these compositions for 28 days, at a temperature of (20±2)°C, until the required resistances are obtained, the mechanical characteristics will be determined [3].

2.2.2. Bending behavior of beams determination. Once poured, dried, and removed from the formwork the small beams reinforced transversely with OB37 stirrups and longitudinally with PC52 can be subjected to mechanical tests to determine the strength and bearing capacity of each element made with each concrete recipe with the addition of recycled materials [14,15].

In order to be able to simulate the deformation, the mode of yielding and breaking of the beams with the help of calculation programs with numerical methods, at the time of the tests, electronic witnesses will be installed to record the displacements.

Thus, the first type of concrete tested in bending on a reinforced beam was **BFAFM** – concrete with fly ash used in a proportion of 10% of the cement dosage and metal fibers with a length of 50 mm used in a proportion of 1% of the mix weight.

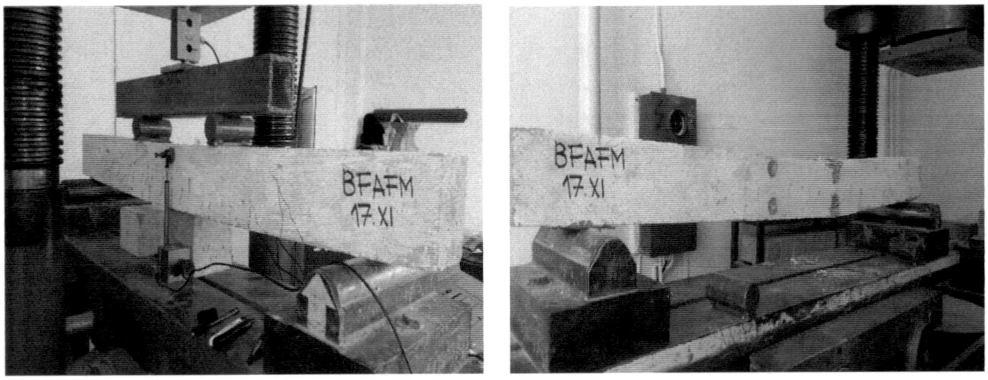

Figure 3. Bending testing of the samples **BFAFM** simply supported beam loaded with two symmetrically arranged concentrated forces.

The second type of concrete tested in bending on the reinforced beam was **BCPES** – concrete with fly ash used in a proportion of 10% of the cement dosage and PES fibers with a length of 30 mm used in a proportion of 0.25%.

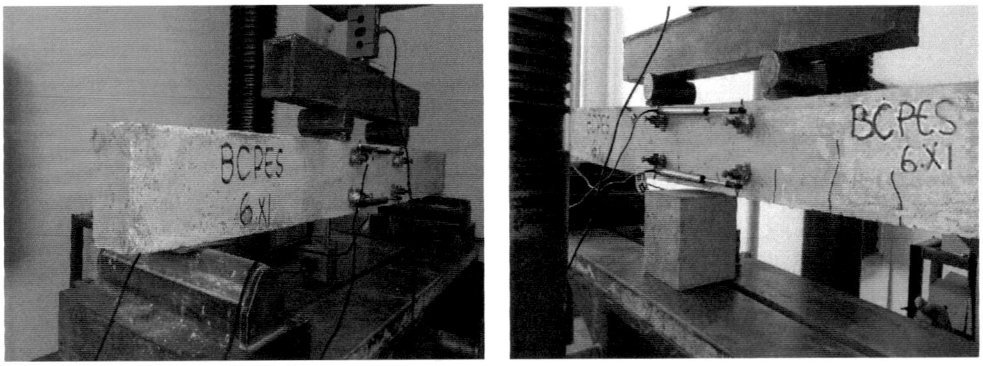

Figure 4. Bending testing of the samples **BCPES** simply supported beam loaded with two symmetrically arranged concentrated forces.

The third type of concrete tested in bending on the reinforced beam was **BFC** – concrete with fly ash used in a proportion of 10% of the cement dosage and hemp fibers with a length of 30 mm used in a proportion of 0.25%.

Figure 5. Bending testing of the samples **BFC** simply supported beam loaded with two symmetrically arranged concentrated forces.

3. Results and discussion

3.1. Influence of recycled materials on Framework, reinforcement and casting of reduced scale reinforced concrete beams

- From the point of view of the execution technology and the way of making the small-sized beams made of concrete with waste additions, no difficulties were encountered, they were made in the same way as well as those made of ordinary concrete.
- Although during the mixing of the component materials, the fibers of PES, tended to form lumps, making it difficult to disperse uniformly in the mass of the composite, at the moment of casting into the mold and putting the material into work, they dispersed, the finished products being approximately homogeneous.
- The vibration of the concrete poured into the beams was done manually, the weight of a beam with formwork and reinforcement exceeding the maximum capacity of the vibrating table.

3.2. Bending tests on concrete elements with additions of recycled materials

- The maximum force applied on the beam made of **BFAFM** – concrete with fly ash used as a substitute in a proportion of 10% of the cement dosage and metal fibers with a length of 50 mm used in a proportion of 1% of the weight of the mix with dimensions 1500 x 120 x 90 mm, was 24200 N = 24.2 kN.

- The failure of the beam occurred through the opening of cracks in the stretched zone and the degradation of the concrete in the compressed zone. The cracks in the stretched area were predominantly vertical and grouped in the bending moment area, which indicates bending failure, the stretching efforts from the shear force being better taken over by the stirrups and the dispersed metal reinforcement. The cracks are developed much vertically, the compression zone being reduced in height.
- The maximum force applied on the beam made of **BCPES** – concrete with fly ash used in a proportion of 10% used as a substitute of the cement dosage and pes fibers with a length of 30 mm used in a proportion of 0.25% with dimensions of 1500 x 120 x 90 mm was 23600 N = 23.6 kN.
- The bending failure of the beam occurred through the development of cracks in the tension zone and the degradation of the concrete in the compressed zone. The cracks in the extended area are predominantly vertical, towards the bearing areas they have a slight inclination, resulting in the failure occurring due to the combined effect of bending moment and shear force. In the case of dispersed reinforcement with PES fibers, the effect of these fibers is no longer as significant as in the case of metallic fibers. The cracks in the central area are smaller in height compared to those in the metal fiber beam. The compressed area is more degraded and develops on a larger surface of the beam.
- The maximum force applied on the beam made of **BFC** – concrete with fly ash used in a proportion of 10% used as a substitute of the cement dosage and hemp fibers with a length of 30 mm used in a proportion of 0.25% with dimensions of 1500 x 120 x 90 mm was 23900 N = 23.9 kN.
- The failure of the beam occurred through the opening of cracks in the stretched zone and the degradation of the concrete in the compressed zone. The cracks in the extended area are predominantly vertical, towards the bearing areas they have a slight inclination, resulting in the failure occurring due to the combined effect of bending moment and shear force.
- In order to be able to establish the qualities and to outline a clear field of use of the elements made of concrete with recycled materials, tests are necessary on all types of concrete studied in each of the forms used.

4. Conclusions

There is a pressing need for sustainable construction practices, as natural resources are depleting and the cost of materials is increasing. The use of waste materials in construction through the adoption of eco-concrete technology can address this issue and contribute to a more sustainable build environment.

Eco-concrete has the potential to reduce the environment impact of the concrete industry by a reduction in carbon emissions, additionally the use of waste materials can increase the industry's use of waste further contributing to a circular economy.

By adopting eco-concrete technology, the construction industry can achieve sustainable construction practices while also reducing environmental problems and protecting natural resources for future generations.

The tests carried out so far have highlighted the fact that some concrete recipes with wastes have obtained results comparable to those of ordinary concretes and deserve further study in order to make the recipe and the production technology more efficient. In the continuation of the research, tests will be carried out on other types of elements (such as blocks, masonry blocks and beams) made with other types of concrete with wastes.

References

[1] Md. Jahidul Islam, Md. Salamah Meherier and A.K.M. Rakinul 2016 *Effects of waste PET as coarse aggregate on the fresh and harden properties of concrete* Construction and Building

Materials **125** 946–95

[2] Elie Awwada, Mounir Mabsout, Bilal Hamad, Mohamad Talal Farran and Helmi Khatib 2012 *Studies on fiber-reinforced concrete using industrial hemp fibers* Constr Build Mater **35** 710–717

[3] Nhamo Chaukuraa, Willis Gwenzib, Tavengwa Bunhuc, Deborah T. and Ruziwad 2016 *Innocent Pumuree Potential uses and value-added products derived from waste polystyrene in developing countries: A review* Resources, Conservation and Recycling **107** 157–165

[4] Ioniuc, I., Diaconu, L.I., et.al. 2016 *The flying ash taken by wet process from cet holboca Iassy–major pollution source* The 9th International Conference INTER-ENG 2015Procedia Technology **22** pp. 413-418 Tîrgu Mureş, România

[5] Alexandru Timu and Marinela Barbuta 2018 *Green concrete with fly ash and plastic waste* Buletin UTI **2**

[6] Nitendra Palankar, A.U. Ravi Shankar and B.M. Mithun 2015 *Studies on eco-friendly concrete incorporating industrial waste as aggregates* International Journal of Sustainable Built Environment **4** pp.378-390

[7] S.C. Boobalan, M. Salman Shereef, P. Saravanaboopathi and K. Siranjeevi 2022 *Studies on green concrete – A review* Materialstoday **65** pp. 1404-1409

[8] Valliyappan T, Siva C, Shanmuga Priyan B, Meganathan K and Manoj Kumar V 2019 *Experimental study on eco- friendly concrete* International Journal of Scientific & Engineering Research **10**

[9] Bărbuta, M., Bucur, R.,et.al. 2016 *Combined effect of fly ash and fibers on properties of cement concrete* 9th International Conference Interdisciplinarity in engineering Procedia Technology **181** Târgu Mureş, Romania, (2016)

[10] Barbuta M.,R. D. Bucur, S. M. Cimpeanu, G. Paraschiv and D. Bucur-Agroecology 2015 Chapter 3, Wastes in Building Materials Industry, INTECH, ISBN 978-953-51-2130-5, pp: 81-99

[11] SR EN 12390-4: Testing Hardened Concrete Part 4 - Compressive Strength - Specification For Testing Machines

[12] SR EN 1992-1-1:2004 – Eurocod 2: Design of concrete structures-Part 1-1: General rules and rules for buildings

[13] SR EN 12390-2: Testing hardened concrete- Part 2:Making and curing specimens for strenght tests

[14] SR EN 3518: Testes on concretes. Determination of the freeze-thawing resistance by measyring the variation of the resistance strength and/or of dynamic relative elastic modulus

[15] SR EN 12390-4: Testing Hardened Concrete Part 4 - Compressive Strength - Specification For Testing Machines

IOP Conf. Series: Materials Science and Engineering 1283 (2023) 012008 doi:10.1088/1757-899X/1283/1/012008

Flood hazard assessment in Montenegro through the application of two-dimensional hydraulic modelling

M Ostojić[1], I Ćipranić[1*], M Jevrić[1], G Sekulić[1] and A Medojević[2]

[1] Faculty of Civil Engineering, University of Montenegro, Montenegro
[2] CEMA ltd, Montenegro

* ivanac@ucg.ac.me

Abstract. Floods, induced partly by climate change and partly by the negative impact of human actions, may be regarded as the most common type of natural catastrophe in Montenegro, causing significant economic, ecological, social, and health damage yearly. Therefore, the focal point of flood risk assessment is reflected in properly determining flood hazards. However, within the Montenegrin National flood risk assessment, which is entirely scenario-based, the hazard is treated only by estimating the occurrence probability of a specific scenario. The paper provides an overview of recent efforts to overcome the potential weaknesses of scenario-based risk assessment by constructing hazard maps for all areas of potential significant flood risk in Montenegro. Hazard maps are intended to visualise different flood intensities (e.g., flood depth, velocity, etc.) for selected return periods. Such outputs are generated by two-dimensional hydraulic modelling, which requires extensive work on gathering the available field data, calculating hydrological inputs, building and calibrating the model, and defining uncertainties that may affect the results. The final production of the hazard maps is conveyed by post-processing the computation results in GIS-based software.

1. Introduction

Permanent or periodic flooding and high groundwater levels endanger over 26,000 ha of land in Montenegro [1]. Flooding in Montenegro is manifested differently depending on the features of the watercourse that causes it. Settlements, traffic infrastructure, industrial facilities, and agricultural areas are commonly located in the downstream valleys of rivers. Their upstream sections are distinguished by significant slopes, high speed of flood wave propagation, and significant quantities of suspended and dragged sediment. Although relatively modest, agricultural areas in these valleys are substantial for agricultural production because the total agricultural resources in Montenegro are pretty limited. Due to this concentration of goods in the river basins, potential damages caused by floods, even on a relatively small scale, can be significant [1].

In addition, uncontrolled urbanisation, especially in the coastal region, contributes to a higher risk of urban flash floods during high-intensity rain events. Also, there is a strong link between urbanisation and flood risk management [2]. The effect of floods in Montenegro is further amplified due to the lack of storm-water sewer systems (or their insufficient capacity), illegal construction work in the floodplains, inefficient river regulation measures, inadequate waste disposal, etc.

The community must make a concerted effort to significantly impact flood risk awareness and local community involvement [3].

Many historic floods that occurred in the past have had catastrophic consequences in terms of damages. The most severe recent events include floods from 2010, which resulted in over 18 million € of injury, as estimated in the damage assessment report composed by the appointed national commission (based on the documentation supplied by local authorities) [4].

The flood risk assessment has yet to be studied for the area of Montenegro. The contribution of this research is reflected in the application of methodology based on risk matrices. Such a methodology allows for considering all the specifics related to flood risk assessment in Montenegro.

2. Flood hazard legal framework in Montenegro – current state

Given the importance of flood-related issues for the future development of the entire water sector in Montenegro, various documents that deal with this matter (at least to a certain degree) have been implemented in the Montenegrin legislative system: National flood protection and rescue plan, National emergency strategy, Water management strategy, Disaster risk reduction strategy, etc.

As a candidate country for joining the European Union, Montenegro was obliged to assess natural and artificial risks and submit data on risk threats to the European Commission per the European Guidelines. As a result, the European Commission approved financing for the project "Disaster Risk Assessment of Montenegro" [5]. As a result of the project, which ended in December 2021, National Disaster Risk Assessment (NDRA) was constituted.

2.1. National disaster (flood) risk assessment

NDRA comprises several independent parts, each dealing with different risk types (floods being one). Flood risk, which according to the EU guidelines may be regarded as a combination of hazard probability and its consequences [6], was calculated only for the selected relevant scenarios - historic events chosen as representative examples for risk assessment purposes. Considering that the main prerequisites for scenario selection are its probability and consequence severity, two types of scenarios were considered:

- the most likely adverse event,
- the event with the worst possible consequences.

Since the entire Montenegrin hydrographic network is almost equally divided between two basins (the Danube basin covering 52.5% and the Adriatic basin covering 47.5% of the territory [4]), each of the two basins is featured through analyses of both scenario types, yielding the total of four flood risk scenarios for the whole country.

The event probability and consequences (on public health, economy/ecology and social stability) were evaluated and combined in a risk matrix (figure 1) for each scenario.

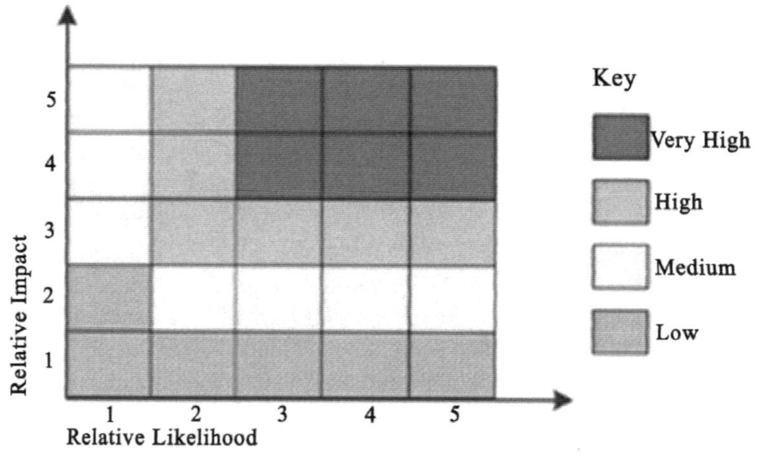

Figure 1. Definition of a Risk matrix for NDRA [6].

For example, floods that occurred in 2016. In the river basin, Lim was chosen as the most likely adverse event scenario for the Danube basin. The hazard return period (15 years in this particular case, which corresponds to a medium probability level) was estimated based on the available historical data by determining the frequency of registered similar events in this basin that occurred in the past. It is worth mentioning that the need for precise, systematic data on past flood events, their causes, and damages makes a historical review of them difficult. Taking the consequences and damages into account requires using not only all available formal data (government reports, local municipality records, etc.) but also the information from the media (papers, television, websites, etc.). When combined in a matrix, the medium likelihood (level 3) and the low consequence level (level 2) generated a moderate risk level.

For NDRA, no hazard maps were made since the hazard intensities were not computed. The only way the obtained results (risk levels) were visualised was by showing the estimated flood extent on a map, uniformly coloured according to the calculated risk level. The boundaries of the floodplain are an approximation based on the field data – locations of damaged households and evacuated people; and not the results of any hydraulic computation.

2.2. *Support for implementation and monitoring of water management in Montenegro*

Most recently, the Montenegrin government has procured funding from the EU (Instrument for Pre-Accession Assistance) to realise the project "Support to Implementation and Monitoring of Water Management in Montenegro." This project aims to support the national and local institutions in aligning with and implementing the EU legislation on environmental protection and climate change [7]. The critical activity in this project includes the preparation of flood hazard and flood risk maps, as well as flood risk management plans to reduce the risk of flood damage.

During the first stage of the project, a Preliminary flood risk assessment (PFRA) was prepared. Historical flood information, geographic data, urban planning information, population statistics, economic activities, digital terrain models (DTM), hydrological and meteorological information, civil protection information, and other national data was used to identify the Areas of Potential Significant Flood Risk (APSFR), which are the priority areas for subsequent further analyses [8]. The second step in the project includes the development of flood hazard and risk maps for all identified APSFRs (25 APSFRs in total were delineated throughout the country), which would later be followed by the third (and final) step – flood risk management planning [9].

The hazard of flooding is distinguished by the probability of its occurrence and the flood intensity (flood depth and velocity). Content of flood hazard maps follows the EU floods directive, covering the geographical areas of all APSFRs and showing the flood extent, water depths and flow velocities for three different scenarios: floods with a low probability (extreme event scenarios), floods with a medium chance (likely return period $\geq$ 100 years) and floods with a high probability [10].

The flood risk, defined as a combination of hazard and damage potential, is assessed by interacting these two factors in a risk matrix. The primary numerical inputs for matrix formulation include the hazard intensity (flood depths and/or velocities) and vulnerability values of assets in the floodplains [11].

Table 1. Risk matrix definition [11].

		Vulnerability value		
		low	medium	high
Intensity	> 5 m or > 2 m/s	moderate (2)	high (3)	high (3)
	1 m – 5 m	moderate (2)	moderate (2)	high (3)
	0.5 m – 1 m	minor (1)	moderate (2)	moderate (2)
	0 m - 0.5 m	minor (1)	minor (1)	moderate (2)

To apply the mentioned methodology for flood hazard mapping and risk assessment, it was necessary to perform the hydrological and hydraulic computation to acquire the flood extent, depths, and flow velocities.

3. Materials and methods

3.1. Study area

Hydrodynamic modelling includes the evaluation of flow characteristics and geomorphic behaviour of rivers due to natural or artificial conditions. The most important results of the modelling are the prediction of water levels and velocities as time and location-dependent functions. The river Lim was chosen as a study area.

Two-dimensional modelling has many advantages compared to traditional one-dimensional modelling, allowing flow conditions to vary in two dimensions instead of one. Two-dimensional models calculate hydraulic results at locations within a mesh that covers the entire geographic extent of a river and floodplain [12]. Two-dimensional hydraulic models in HEC-RAS [13] were developed for flood hazard mapping for each APSFR in Montenegro.

Here is the presented model and results of the river Lim. These models are depth-averaged, assuming that velocity is averaged in the vertical (z) dimension and varies in the longitudinal and lateral (x,y) directions.

3.2. Data sets

Obtaining hydrological data and terrain data inputs was necessary to realise this research.

3.2.1. Terrain data inputs. Terrain data (for flood hazard purposes – hydrology and hydraulic analysis) is one of the essential input datasets for modelling. The quality of model results and mainly the following data processing depend significantly on the quality of the terrain data.

The terrain data was provided as Digital Elevation Models (DEMs) by the Montenegrin Real Estate Administration, created by LiDAR (Light detection and ranging) technology [14]. The resolution of the available elevation models is 5 m x 5 m, which, to a certain degree, satisfies the recommended resolution requirements of terrain data for flood risk management issues. However, this data should be combined with geodetic survey data (terrestrial and bathymetrical) to increase the accuracy of the details concerning buildings and other terrain structures near water bodies [11]. Unfortunately, no surveyed cross-sections were available for the mentioned project. No further surveying campaigns were feasible, considering the size of the study areas and limitations regarding the project timeframe and available funding. One of the future recommendations for improving the models would include surveying all buildings in and around the water bodies. Also, river cross sections, especially those occupied by water during the LiDAR survey, should be surveyed to build a stable model setting for hydraulic flow properties.

3.2.2. Hydrology inputs. The available measured hydrological data and existing historical data (from recorded historical events) were supplied by the Institute of Hydro-Meteorology and Seismology of Montenegro (IHMS). For each APSFR, datasets (water elevation, flow rate, precipitation, etc.) from the nearest measuring station were statistically processed. Parametric hydrology methods were applied to estimate the relevant flow rates and hydrographs for each selected return period: 10, 100 and 500-year events. These were then used as boundary conditions for the hydraulic models.

Hydraulic models are susceptible to roughness values (Manning's *n*). Hence, Manning's *n* needs to be chosen carefully and calibrated if possible (when discharge and flood extent of historic floods is available). The importance of Manning's n values for different land use classes increases with extensive flood plains. Other literature sources and recommendations were used to find an appropriate Manning's n value [12].

4. Results

Unsteady conditions for flow modelling were applied (considering the characteristics of the study areas and the common flood types) since an unsteady flood model can better display the dynamic scenarios. In this manner, the time-dependent flow was considered, due to which the system changes over time. When hydrograph data (from recorded historical flood events) was unavailable, synthetic flow hydrographs derived from parametric hydrology methods were used.

Three simulations were conducted for every study area, each one corresponding to different hydrological conditions in terms of occurrence probability: 10-year (10% probability), 100-year (1% probability), and 500-year (0.2% probability) event. Each analysed area's simulation duration was selected to depict the flood wave propagation best, while the Courant condition controls the time step.

Results of the simulation – flood extent and water depths (HQ10, HQ100, and HQ500) for the river Lim, which was one of the designated APSFRs, are shown in figure 2.

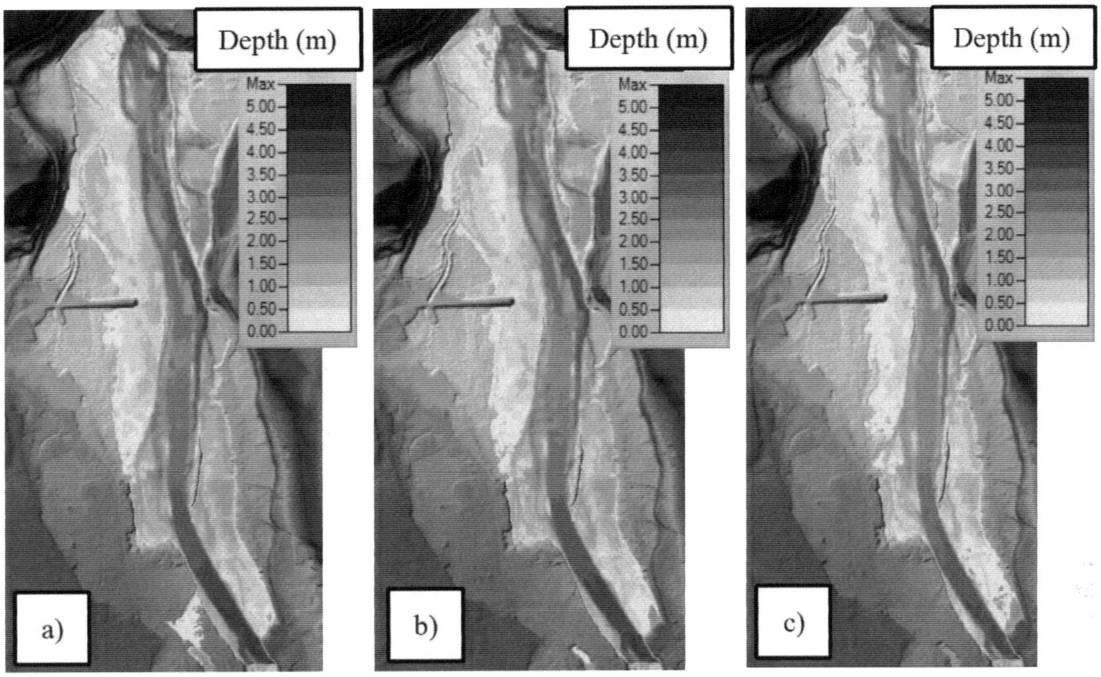

Figure 2. Flood extents for return periods HQ10 (a), HQ100 (b), and HQ500 (c).

After modelling, the results must be processed for all defined scenarios. The datasets required for further processing for hazard map production are flood extents, inundation depth, and flow velocities. Finally, the model outputs are exported as shapefiles and used for map building in GIS-based software.

Two different types of hazard maps are constructed:

- Flood Hazard Map shows the flood extents of the different return periods. Three layers representing flood extents for all three return periods are overlapped on the map, showing buildings, bridges, and river channels displayed over an orthophoto background (figure 3).
- Flood Hazard Map for each return period showing the inundation depths. For all three maps (each referring to one return period), calculated depths are classified into three ranges: 0 - 0.5 m, 0.5 - 1.5 m, and > 1.5 m, and shown as different layers. The flow area with velocities greater than 1m/s is also delineated on the map, along with buildings, bridges, and river channels displayed over an orthophoto background (figure 4).

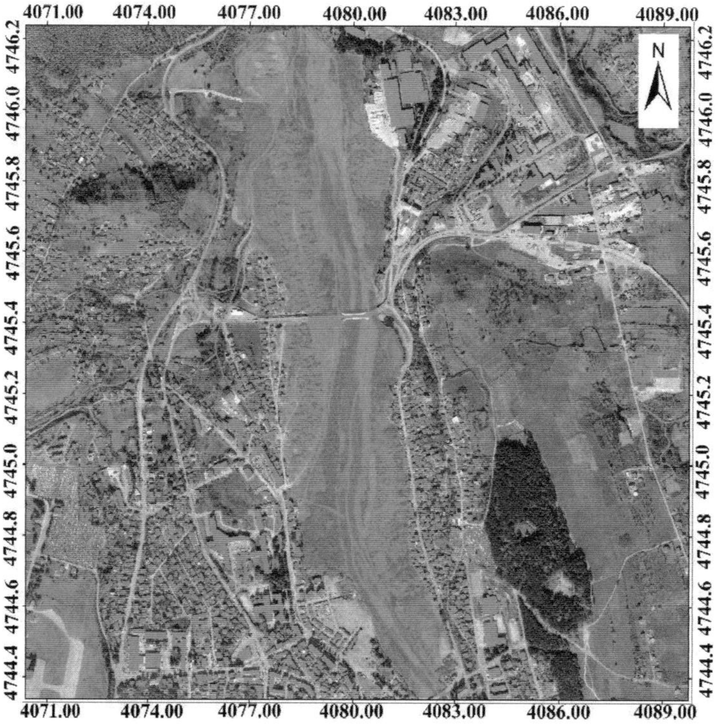

Figure 3. Flood hazard map with flood extents for all return periods – example of river Lim.

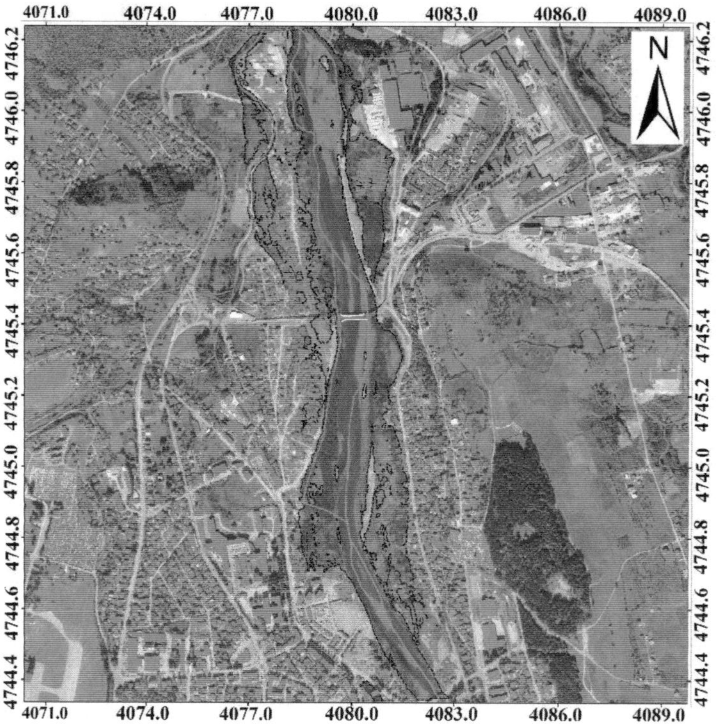

Figure 4. Flood hazard map with inundation depths for HQ500 – example of river Lim.

The computed inundation depths from the model also represent a basis for further risk assessment. According to the suggested risk-matrix method (table 1), the risk is calculated by combining flood exposure and vulnerability. The first factor, flood exposure, is defined using intensity classes showing different inundation depths. Therefore, computed depths must be classified according to table 2 and exported to formats appropriate for subsequent risk calculation.

Table 2. Classification of intensities of inundation depths (for the definition of flood exposure) [11].

Water depth	Class
< 0.5 m	1
0.5 - 1 m	2
1 - 5 m	3
> 5 m	4

5. Conclusion

Assessing and reducing the risk of floods in the entire country represents a permanent task by improving the protection of the most critical damage centres (cities, populated areas, business entities, traffic infrastructure, etc.), and regulation works and measures on watercourses represent priority activities. Therefore, flood hazard and risk maps are important information bases for different user groups to reduce the risk and damage caused by flood events. Based on the information in the maps, users integrate hazard and risk information into their decision-making, planning, and instruments [11].

The National flood risk assessment, as the principal legal document on this matter, offers a scenario-based methodology for determining the levels of risk. The most prominent disadvantage of such an approach is using a few arbitrarily selected scenarios (which rely on historical flood data) to depict the flood risk for the whole country's territory. This may lead to overlooking critical scenarios – catastrophes can be neglected, and at the same time – low probability-high consequence events are classified as medium-risk events by definition (table 1). Montenegrin NRA does not include any hazard maps. In addition to this, risk maps (existing only for the selected scenarios) show uniform risk descriptions for the entire scenario-designated area, regardless of the position of risk assets in the floodplain (dedicating the same level of risk for the objects at the very border of the floodplain, and the objects that may be located in the river channel itself, which are exposed to much greater flood depths and velocities).

However, the water management strategy in Montenegro recognises projects whose implementation would have an extremely significant positive effect in protecting against floods. One of them is *Support for implementing and monitoring water management in Montenegro*. Therefore, flood hazard map production for all APSFRs in Montenegro is among the critical products expected upon the realisation of this ongoing project. They are meant to visualise essential parameters describing the intensity and character of a flood:

- The flood extent – shows the inundated area for different flood scenarios.
- The inundation depth – indicates, for example, up to which height a building will be flooded and on which floors assets and persons will not be affected by the flood.
- Velocity – describes how fast the water is moving. Depending on the flow velocity, the risk of adverse consequences to human health, the environment, cultural heritage, and economic activity might increase. At high velocities, people and objects are more likely affected by floods.

These parameters could only be obtained through hydraulic computation of the study areas. For these purposes, two-dimensional hydraulic models were developed because complex flow situations (as with extensive flood plains, levees, multiple directional flows, etc.) are handled more suitably with the two-dimensional approach. In addition, two-dimensional models provide a more intuitive graphical representation of results than cross-sections and tables from a one-dimensional model [12]. However, preparing the inputs for the model (especially terrain data) should be done with great care since the reliability of the results depends significantly on the quality of the input data. There is always room for improvement of the newly developed models (when the time comes to update the maps), particularly regarding additional geodetic surveying, which would further enhance and improve the used DEMs.

References

[1] Government of Montenegro, Ministry of Internal Affairs, Directorate for Protection and Rescue 2021 Disaster Risk Assessment of Montenegro (Podgorica)
[2] Shafira S 2022 Analysis of the correlation urban systems and risk management floods, study case: Kaliwungu, Kendal Regency IOP Conf. Ser.: Earth Environ. Sci. 1082 012025
[3] Sedyowati L, Chandrarin G and Nugraha G I K 2021 Towards sustainable community: Effort to reduce flood risk and increase well-being in a densely populated flood-prone area IOP Conf. Ser.: Earth Environ. Sci. 930 012096
[4] Government of Montenegro 2019 National flood protection and rescue plan (Podgorica)
[5] Government of Montenegro, Ministry of Internal Affairs, Directorate for Emergency Situations 2020 Instructions for preparing a disaster risk assessment in Montenegro (Podgorica)
[6] European Commission 2010 Risk Assessment and Mapping Guidelines for Disaster Management (Brussels)
[7] https://webgate.ec.europa.eu/europeaid/online-services/index.cfm?ADSSChck=1666103353578&do=publi.welcome&orderby=upd&orderbyad=Desc&aofr=139429&nbPubliList=15&searchtype=RS&userlanguage=en March 1, 2023
[8] Support to implementation and monitoring of water management in Montenegro, EuropeAid/139429/IH/SER/ME 2021 Preliminary Flood Risk Assessment for the Danube River Basin (Podgorica)
[9] Support to implementation and monitoring of water management in Montenegro, EuropeAid/139429/IH/SER/ME 2021 Preliminary Flood Risk Assessment for the Adriatic River Basin (Podgorica)
[10] Official Journal of the European Union 2007 Directive 2007/60/EC of the European Parliament and of the Council of 23 October 2007 on the assessment and management of flood risks
[11] Deutsche Gesellschaft fur Internationale Zusammenarbeit (GIZ) GmbH 2022 Flood Hazard and Risk Mapping for the Drin/Drim – Buna/Bojana River Basin Guidebook (Darmstadt / Heidelberg / Tirana)
[12] U.S. Department of Transportation, Federal Highway Administration 2019 Two-dimensional Hydraulic Modeling for Highways in the River Environment Reference Document
[13] https://www.hec.usace.army.mil/software/hec-ras, March 4, 2023
[14] https://lidarradar.com/apps/100-applications-or-uses-of-lidar-technology, March 8, 2023

Analysis of mortars with old plastering waste using state-of-the-art method: nuclear magnetic resonance

C Saitis[1], D L Manea[1], R Fechete[2], L M Plesa[1] and L Saitis[2*]

[1] Technical University of Cluj-Napoca, Faculty of Civil Engineering, Cluj-Napoca, Romania
[2] Babes Bolyai University, Faculty of Physics, Cluj-Napoca, Romania

* saitis@mail.utcluj.ro

Abstract. One of the major problems we face today is the correct management of waste generated in all spheres of activity. Thus, the concept of sustainable development aims to develop strategies that prevent and solve problems already generated. Domain construction is the one that generates the most waste worldwide, according to recent studies, which requires the imposition of clear, effective, and future-proof measures to improve the environment and the health of the population. This paper is the continuation of a research program in which 3 types of mortar recipes are analyzed. The study involves replacement of aggregates, in different percentages (10% and 15%), with plaster waste from historic buildings. The obtained results show that the values fall within the limits stipulated in the norms in force, so the study will continue with the analysis of the samples using the state-of-the-art method: nuclear magnetic resonance. The paper aims to analyze the properties of the proposed mortars at the nanoscopic level, following the relationship between the plaster mortar matrix and the plaster waste introduced in different percentages in the samples. In addition, the impact of these wastes on the hydration of the cement and on the porosity of the composition of these mortars is also observed, being noticed that the relationship between the porosity of the samples and the mechanical resistances is inversely proportional, that is, the resistances decrease as the porosity increases. It will also be identified the location of hydrogen in the structure: in hydrated mineralogical compounds of cement and in large, medium and small pores. For each studied sample, the investigations were conducted at the age of 7, 14 and 28 days (even at 3 days for samples with 10% and 15% substitution of aggregates with old plastering mortars). The results for the proposed samples will be compared to the standard recipe.

1. Introduction

In 1970, the concept of sustainable development was established because of the effects that pollution had since then on the environment and on the health of the population. Thus, over time, short, medium, and long-term strategies have been developed to reduce the problems already encountered and to prevent the occurrence of other negative effects on the environment. These strategies have also been introduced in the construction sector, which according to Eurostat studies is the sector that generates the most waste with a contribution of 807.372 million of tons (37.5% of waste production in 2020) [1]. One of the many methods that can help to reuse the generated waste is their inclusion in the mass of mortars by displacement of components in different percentages, so that the requirements for the minimum physical-mechanical characteristics stipulated in the regulations are further met [2].

Nuclear magnetic resonance (NMR) is a powerful method used to study materials from elastomers, gels, biomaterials, and porous materials. Thus, much of the processing of data related to NMR measurements for porous materials was developed by oil companies interested in the study of pores, water in these pores, the distribution of pores and especially crude oil in pores as large as possible and as well connected as possible. One of the areas of use of specific porous materials is that of construction, where many types of mortars are used.

Thus, nuclear magnetic resonance, especially in low fields, can be used to measure distributions of spin-spin relaxation times, mainly for water that is in various types of pores, from large, medium pores, small pores but also as hydration water in the components of the mortar. By a low-field NMR measurement, a nuclear magnetization drop curve is obtained which is considered multi-exponential and which is analyzed by the reverse Laplace transform, where it is usually obtained a distribution of T2 spin-spin relaxation times that give us information about pore size and distribution. So, the water molecules in the pores can diffuse, but only within the limits imposed by their size, and the relaxation time T2 is sensitive to this limit, so for free water it can be obtained a T2 relaxation time somewhere between 2.4 and 3 seconds, and as the size of the pores leads to a restriction of water diffusion, and water molecules collide with the walls of the pores, the T2 relaxation time has increasingly smaller values.

Considering the numerous studies in this domain, it can be mentioned that there is a considerable diversity of waste that can be reused in the composition of concrete/mortars, since in most cases, both in the substitution of cement and in the substitution of aggregates, the values obtained meet the conditions stipulated in the regulations, so this paper addresses new ways of re-using waste from the point of view of sustainable development [3-11].

For instance, the nanoscopic analysis was used by Pintea et al [12] in discovering that casein and rice if added in small percentages (2%) to plastering mortars could lead to an increased homogeneity of the water media and the pore structures, whereas if added in higher percentages (10%) resulted in less homogeneity of the samples. It was also shown in the study that in the case of dry sample the small and medium pores are partially saturated and partially homogenous.

Jumate et al. [12] also used the NMR method in studying the effect of addition of cellulose ether polymer and $CaCO_3$ filler in plastering mortars. The results showed that these components induced a larger distribution of pores and channels, a decrease of small pores and an increase of large pores at the same time [13].

The present research is in fact the continuation of a study presented at a conference in Brno, Czech Republic in November 2022 (Recycling Plaster Waste as a Substitute for Aggregates in Obtaining Plastering Mortars [14]) where the results of physical mechanical strengths were analyzed both in a hardened and in a fresh state for the prepared samples. The main subject of the article was analyzing the physical-mechanical properties of the mortar samples obtained by replacing the aggregates with old plastering mortars in proportions of 10% and 15% at the age of 7, 14 and 28 days.

2. Materials and methods

Plastering mortars waste from old buildings represents a growing problem in terms of sustainable development because of the degradation of these buildings that constantly need to be rehabilitated. The quantity of these kind of waste grows each day, but the composition of those old plastering mortars could be reused in making new types of mortars when added in different percentages in new types of mortars. Reusing these wastes could bring many benefits to the environment especially when there can be involved the saving of raw materials.

The materials used in this study are submitted with requirements set up in regulations SR EN 998-1:2004. Masonry mortars specifications. Part 1: Plastering and glazing mortars, SR EN 998-1/ AC-2006. Masonry mortars specifications. Part 1: Plastering and smoothing mortars and SR EN 1015-2:2005. Masonry mortars testing methods. Part 2: Overall sampling of mortars and preparation of the mortars to be tested [15-17].

2.1. Materials

Cement – for all the samples it was used Portland 52.5 R cement that was fabricated in Alesd city, country of Romania. According to SR EN 197-1:2011, the main characteristics of this material are the following: composition of the cement is composed by 80-94% clinker, 6-20% blast furnace slag, 0-5% minor auxiliary components; minimum 45 minutes plug time; minimum 10 mm expansion' minimum 30 MPa initial compressive strength; minimum 52.5 MPa standard compressive strength; maximum 4% sulphates content; maximum 0.1% chloride content and maximum 0.0002% hexavalent chromium content [14].

Aggregates – sand divided into 3 sorts in size which are 0-0.5 mm, 0.5-1 mm, 1-2 mm. The quantity of sand was equally divided in each sort with 483.33 kg/ m^3.

Lime – hydrated calcic type, Carmeuse Supercalco M, fabricated in Brasov city, Romania.

Old plastering mortars waste – all the waste was collected directly from the site, more specifically from the brick backing layer of a historical building façade that will be rehabilitated. The first phase of the rehabilitation was the undoing of all degraded plastering mortars. The collected samples, according to the building's technical project, are all lime mortars and they were taken to the testing laboratory of building materials of the Technical University of Cluj – Napoca, Faculty of Civil Engineering where they were used as replacement for aggregates in two different percentages: 10% and 15%, respectively. In figure 1 it can be observed the brick support layer of the old plastering mortars. The collected samples were free of brick debris.

Figure 1. Brick backing layer for old plastering mortars.

Afterwards, the waste was shredded and divided into 3 sorts in size: 0-0.5 mm (figure 2a), 0.5-1 mm (figure 2b), and 1-2 mm (figure 2c), and then they were used in preparation of mortars by replacing aggregates in the proportions mentioned above.

(a) (b) (c)

Figure 2. The 3 sorts in size of old plastering mortars.

2.2. *Studied recipes*

The 3 types of mortars that are subject to the current study are CS III STANDARD. – a cement – lime mortar without added waste, CS III SUBSTITUTION 10% - a cement lime mortar in which the aggregates were replaced by old plastering wastes in proportion of 10% in each sort in size (0-0.5 mm, 0.5-1 mm, 1-2 mm) and CS III SUBSTITUTION 15% - a cement lime mortar in which the aggregates were replaced by old plastering wastes in proportion of 15% in each sort in size (0-0.5 mm, 0.5-1 mm, 1-2 mm) as shown in table 1:

Table 1. Studied recipes.

Recipe	Waste displacement [%]	Old plastering waste [kg/m^3]	Water [l/ m^3]	Cement [kg/m^3]	Lime [kg/m^3]	Sand [kg/m^3]
CS III STANDARD	0	0	311			1450
CS III SUBSTITUTION 10%	10	145	326	275	110	1305
CS III SUBSTITUTION 15%	15	217.5	331			1232.5

2.3. *Methods*

The samples were prepared in the construction materials laboratory of Technical University of Cluj – Napoca. The preparation of mortars consists of cleaning and drying the aggregates, dividing them into 3 sorts in size, establishing the exact amount of each material needed (cement, aggregates, lime, water) and mixing them together with the mortar mixer until reaching the requested homogenization [14]. Then, the mortar samples were analyzed at the age of 3, 7, 14 and 28 days according to NMR spectroscopy analytical method [18].

NMR method involves the analysis using the reverse Laplace transform which allows us to use an undefined model, that is, we will not assume the number of pores, their distribution and the amount of water bound, because this results naturally from the Laplace analysis. We will also get at low values, usually below a millisecond, a drop that can be associated with water bound to values around a millisecond. At longer times, droplets are associated with water from small pores, and at values of several tens of milliseconds, drops associated with water from medium pores are obtained, and at values of several hundred milliseconds, drops associated with water from large pores are obtained. At intermediate values of relaxation time, if obtained for these intermediate values, the droplets may be associated with water from small to medium pores or medium to large pores. All the measurements analysis had been realized with the NMR apparatus shown in figure 3, in the Physics laboratory of Technical University from Cluj-Napoca [19, 20]

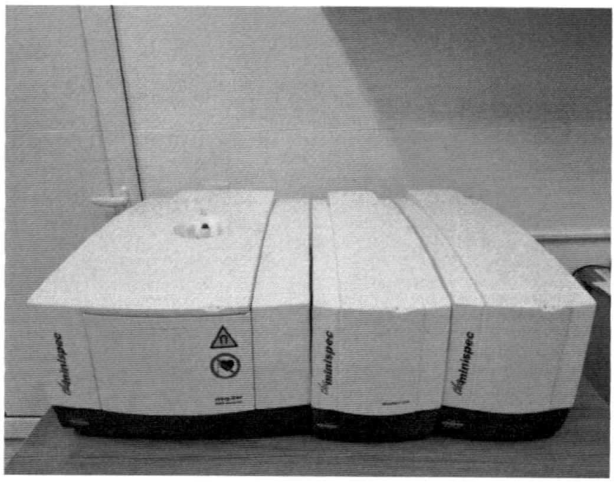

Figure 3. NMR apparatus.

3. Results

The physical-mechanical strengths were all presented at an international scientific conference in Brno city, Czech Republic in November 2022 and this is the reason why the paper focuses on NMR (nuclear magnetic resonance) analysis.

All the results obtained for the fresh state (apparent density, consistency, and segregation tendency) and for the hardened state (apparent density, compressive strength, bending strength, adhesion to the support layer – at the age of 7, 14 and 28 days) met the values stipulated in the European norms in force, even though it could be observed a slight decrease in the results obtained for the analyzed strengths when old plastering mortars waste was added [14, 21, 22].

- CSIII STANDARD samples analysis

Figure 4 shows the distributions of T2 relaxation times measured for CSIII STANDARD mortar with a squeal occurrence time of 70 μs at 7, 14 and 28 days after preparation.

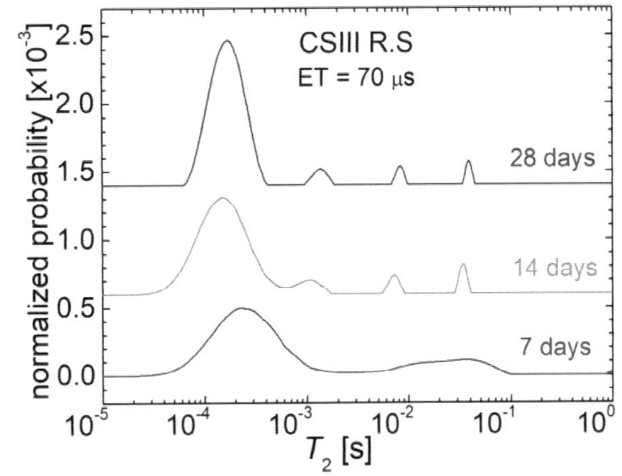

Figure 4. CSIII STANDARD at the age of 7, 14 and 28 days.

It is observed that seven days after preparation a pore appears with a large drop located at about 250 μs, but also with a slightly localized shoulder at about 700-800 μs. This indicates that most of the water is found in the form of water linked to the components of the mortar and that there may also be a small number of small pores. At the age of seven days, two not very well solved drops are observed centered at approximately 10 and 40 Ms, which indicates the formation of medium and medium to large pores. At 14 days it is observed that the pore structure is well defined (according to the orange curve in the figure), so we have a water bound given to a drop that is centered at about 50 μs, a fairly small amount of water located in small pores - the dropper located at about 1 ms and a comparable amount of water in medium and medium to large pores - the two droplets located at about 8 ms and 35 ms.

The fact that they are quite narrow at the base indicates a high homogeneity of the sample. For the CSIII STANDARD mortar, up to 28 days, there is no very large difference in terms of its workability, the only notable difference (between 14 and 28 days) is achieved at low values of T2, where there is a narrowing of the main drip which is associated with the water bound and the appearance of a landing until the next drip associated with the water in the small pores, which suggests that this distribution of T2 indicates that the water bound is now well delimited by the water in the small pores.

- CSIII SUBSTITUTION 10% samples analysis

Figure 5 shows the distributions of T2 relaxation times measured for CSIII SUBSTITUTION 10% (addition of waste from old plastering mortars) with a 70 μs echo occurrence time at 3, 7, 14 and 28 days after preparation.

By adding 10% waste of old plastering mortar that replaced the aggregates, it is noted that as early as 3 days after the preparation of the new mortar a pore structure is formed (green curve figure 5) where the medium and medium to large pores are already well defined, determined by the two droplets located at about 7-8 ms and 35 ms respectively (approximately). And in this case, as in the case of CSIII STANDARD, 7 days after preparation, it is observed that for the water bound, a dropper appears at about 350 μs, but which also has a shoulder centered at about 1 ms and which indicates the formation of small pores.

At 7 days for this mortar sample, a well-formed pore structure is already observed, thus appears: a main dropper located at about 200 Ms and associated with water bound, small pores that are represented by a drop located at about 1.5 ms and medium and medium to large pores whose droplets have moved to slightly higher values of T2, this indicates a slight increase in pores size compared to the distribution of T2 measured at 3 days.

At 14 days after preparation, few changes are observed for this recipe, but 28 days after preparation it is observed that the water in the large pores represented by a drop located at about 30 - 35 ms, it is about the same amount as the water in the large pores from 14 days.

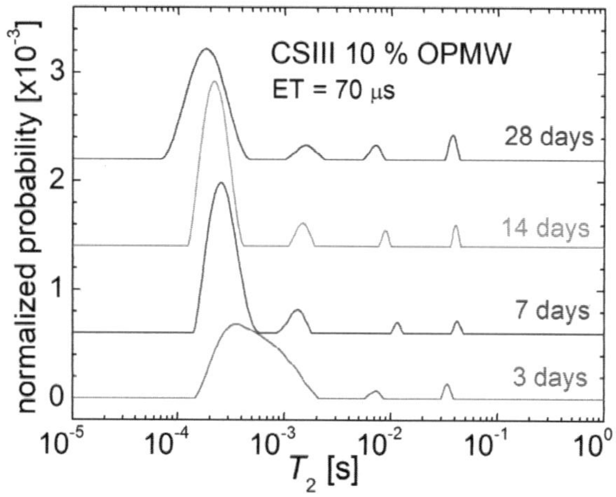

Figure 5. CSIII SUBSTITUTION 10% at the age of 7, 14 and 28 days.

The medium pores have decreased slightly in size, but the relative amount of water seems to be slightly larger, as well as the small pores are also centered at 1-1.5 ms, which proves that the size of the pores for the same mortar at 7 and 14 days has not changed essentially, but the droplets are wider, suggesting greater heterogeneity and inhomogeneity. The same is observed for the drop that is associated with the water bound, becomes wider than the drop measured at 7 and 14 days, which tells us that it is more heterogeneous (non-homogeneous) and localized at slightly lower T2 values, which indicates a slight increase in the compaction of the sample.

- CSIII SUBSTITUTION 15% samples analysis

Figure 6 shows the distributions of T2 relaxation times measured for CSIII SUBSTITUTION 15% (addition of waste from old plastering mortars) with a 70 μs echo occurrence time at 3, 7, 14 and 28 days after preparation.

The mixing of the aggregates with 15% waste of old plastering mortar indicates a clear pore structure from 3 days after preparation. The bound water is represented by a narrow drop located at about 350 μs, while the small pores are represented by a narrow drop located at about 1.5–1.8 ms. So are the medium pores represented by a dropper located at about 17-18 ms and the medium to large pores represented by a droplet located at about 60 ms. The mortar at 7 days shows the same pore structure, only a slight decrease in the value of T2 is observed for which the dropper associated with the bounded water appears, which tells us that the mortar is strengthened a little more. Otherwise, the structure of pores by the positions of the droplets that appear suggests that they are the same, only the amplitude of the medium and medium to large pores is smaller, which probably indicates a slight relative loss of water in these pores, most likely by evaporation.

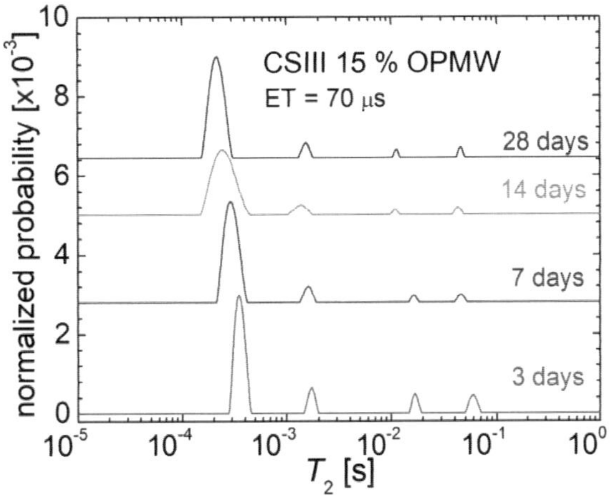

Figure 6. CSIII SUBSTITUTION 15% at the age of 7, 14 and 28 days

The same phenomenon is observed in the 14-day sample, the same pore structure is observed with relatively small amplitudes for water from medium and medium to large pores, slightly moved to lower values, which indicates a slight decrease in their size. Instead, there is a slight increase in the width of the drop associated with the water bound, which indicates an evolution especially in this area of the water bound.

At 28 days, minor changes in the pore structure are observed, only a slight narrowing of the drop associated with the water bound, indicating an increase in the homogeneity of the sample and a slight decrease in the value of T2 at which it occurs at 200 μs. This indicates an increase in the compaction of the sample. In fact, there is a slight decrease in the value of this main dropper associated with water binding for as many as 3 days to 28 days, indicating a slight increase in the compaction of the samples as they evolve over time, i.e., as they hydrate.

To assess as best as possible, the dislocation of aggregates with old plaster mortars, figure 7 shows the measured distributions for T2 relaxation times compared to different samples at the same time after preparation.

- Analysis of the samples at the same age

Figure 7 compares distributions for CSIII STANDARD, CSIII SUBSTITUTION 10% and CSIII SUBSTITUTION 15% mortars where aggregates are gradually replaced in different percentages. Large differences are observed, so if at 3 days CSIII SUBSTITUTION 10% barely shows a pore structure, at values less than 1 ms, we can see a drop that can be associated with water tied and a shoulder that indicates the presence of a structure with small pores. It is also noted that the average pores are very wide by the presence of the dropper at 7-8 ms and the medium to large pores represented by a drop located at about 30-35 ms. Instead, only by adding another 5% waste from old plaster mortars can we see a very clear definition of the porous structure, and the homogeneity of the sample increases rapidly. Thus, we have a narrow drop located at about 400 μs, representative of the water bound, which has an amplitude of 30% of the amplitude of the main drop and which represents a relatively large amount of water in small pores, very well structured. This structure is observed from the fact that the dropper is narrow. So do two more narrow droplets located at about 17-18 ms and about 40-50 ms, which represent the water associated with the medium and medium to large pores.

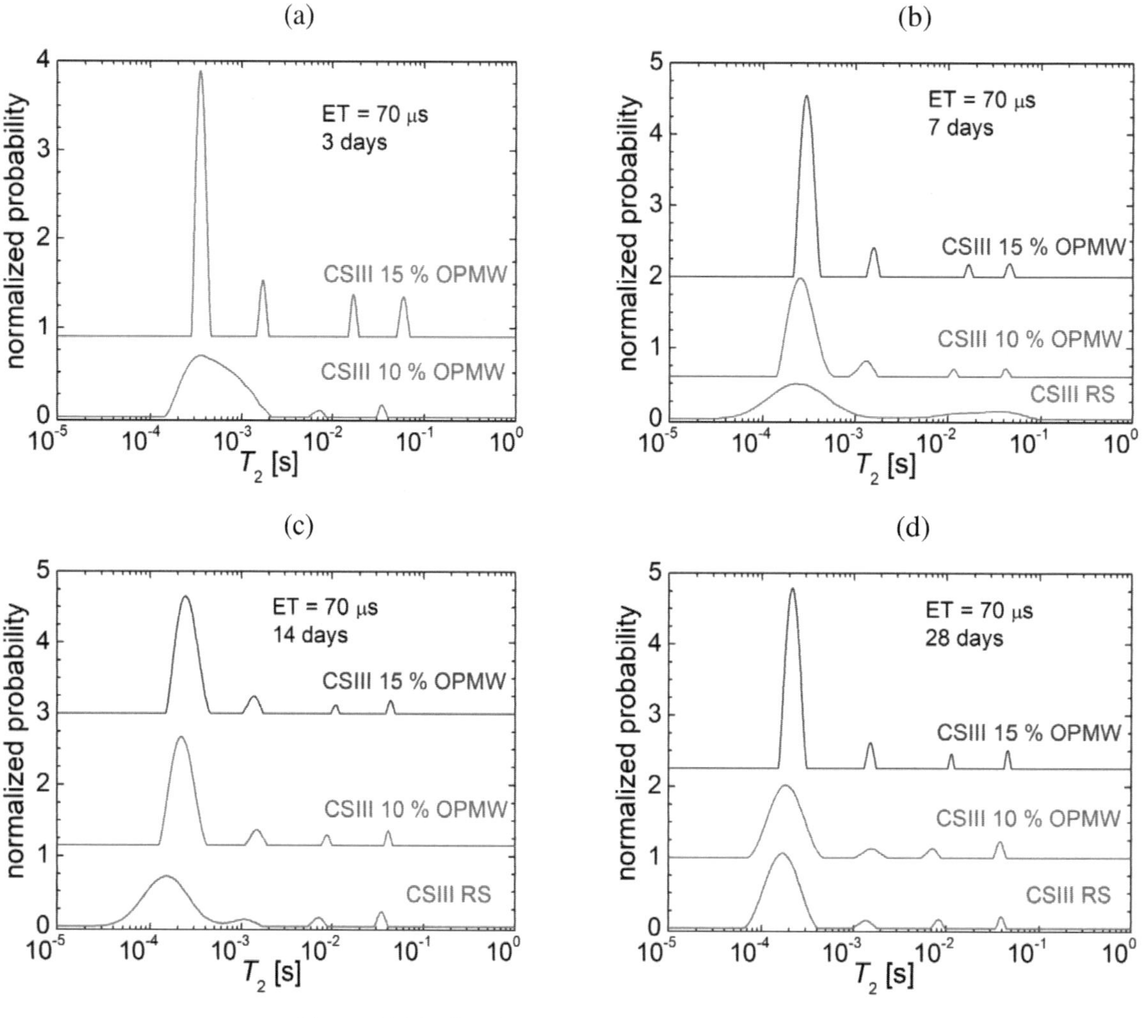

Figure 7. T2 distributions at different time intervals.

At 7 days after preparation, it is noted that the standard CSIII STANDARD mortar shows some large droplets, while by adding 10% waste from old plaster mortars, the porous structure is already present and the hydrogen in the bonded water is already well differentiated from the hydrogen in small pores, medium and large. This differentiation from 7 days after preparation is even better seen when there is 15% waste in the composition.

At 14 days after preparation, it is noted that in all cases, regardless of the amount of waste in the composition, even in the case of standard mortar, a very well-defined pore structure is obtained, even more, for standard mortar the water bound leads to the dynamic homogeneity of the samples of these rigid components. A well-formed pore structure is also observed if we add 10% and 15% waste to the mortar composition. This addition decreases the homogeneity of the mortar structure in terms of the dynamics of the rigid components, therefore the water bound as seen from the drop present at about 200-250 μs. By adding 10% waste it reduced the size of small pores and by adding 15% waste the size of small pores decreased even more, while this decrease from the standard is observed for all types of pores even medium and medium to large at 10%. It is noted that at CSIII SUBSTITUTION 15% compared to CSIII SUBSTITUTION 10% an increase in the size of medium and medium to large pores is obtained. At the same time, by adding 15% waste from old plastering mortars, it seems that water through evaporation is lost in a larger amount.

At 28 days it is noted that the standard mortar CSIII STANDARD and CSIII SUBSTITUTION 15% have a similar pore structure (composition), and very well-defined droplets are observed at about 200 µs for water bound, at about 1,5 ms for water in small pores, about 10 ms for water in medium pores and about 40 ms for water in medium to large pores. In contrast, for CSIII SUBSTITUTION 10%, approximately the same pore structure is observed only that they are less homogeneous, which is observed from the fact that the distributions are wider, and for small pores by a slight displacement to T2 values at about 1,5 ms, a slight increase in the size of small pores is also indicated. For the other two types of medium and medium to large pores, the location of the drops is the same as that discussed for the other two samples, only the droplets are wider, so in this case too, we can talk about a slight inhomogeneity of the samples. It is also noted that shortly after preparation (three days, seven days), the use of waste from old plastering mortars that already have hydrated components, led to an earlier formation of the pore structure than in the case of the standard CSIII STANDARD mortar.

4. Conclusions

The addition of waste from old plastering mortars can influence the characteristics of the analyzed mortars as it follows:

1. At the age of 3 days the pores structure of CSIII SUBSTITUTION 10% recipe could barely be seen whereas the CSIII SUBSTITUTION 15% one was well defined which means a higher homogeneity (for the water media) of the sample.
2. At the age of 7 days the pores structure is better defined, and its homogeneity gets higher as waste is added. The differentiation between the hydrogen in the bonded water and the hydrogen in the small, medium, and large pores could be well observed at this age.
3. At the age of 14 days the homogeneity of mortar structure in terms of rigid components decreases as waste is added. The pores' structure is well defined for each sample and the water evaporation is larger for CSIII SUBSTITUTION 15%.
4. At the age of 28 days the pores' structure is similar for CSIII STANDARD and CSIII SUBSTITUTION 15% whereas for CSIII SUBSTITUTION 10% it is less homogeneous.
5. Well defined pore structures appear at early ages for samples with old plastering mortars waste because this material already has in its composition hydrated components.
6. The fact that the relaxation time distributions for CSIII STANDARD and CSIII SUBSTITUTION 15% are very similar, indicates that this amount of 15% waste from old plaster mortars can be successfully used in tested mortars recipe.
7. It is proposed to continue the research with the following procedures: displacement of aggregates with old plastering in a higher proportion than 15%; continuation of NMR analysis with the deconvolutions related to saturated samples and testing of mortars with SEM (electronic sweep microscopy) and X-ray diffraction.

The potential of this substitution was shown by the values obtained which were still higher than the ones stipulated in the norms in force. Taking into consideration the fact that facades rehabilitation generates large amounts of waste it is desirable to reuse as much quantity of old plastering mortars as possible.

References

[1] https://ec.europa.eu/eurostat/statistics-explained/index.php?title=Waste_statistics#Total_waste_generation, accessed October 10, 2022
[2] *Reducing the impact of construction on the environment* – ISO press release no. 1344/ 2010
[3] Kajanan S et al. 2021 *Development of sustainable mortar using waste rice husk ash from rice mill plant: Physical and thermal properties* Journal of Building Engineering **43**
[4] Luhar S, Cheng T-W, Luhar I 2019 *Incorporation of natural waste from agricultural and aquacultural farming as supplementary materials with green concrete* Composites Part B **175** 107076

[5] Sandanayake M, Bouras Y, Vrcelj Z 2022 *A feasibility study of using coffee cup waste as a building material - Life cycle assessment and multi-objective optimisation* Journal of Cleaner Production **339** 130498
[6] Dawood E T, Shawkat A S, Abdullah M H 2021 *Flexural performance of ferrocement based on sustainable high-performance mortar* Case Studies in Construction Materials **15** e00566
[7] Al-Akhras N M, Abu-Alfoul B A 2002 *Effect of wheat straw ash on mechanical properties of autoclaved mortar* Cement and Concrete Research **32** 859–863
[8] Molnar Luminita Monica. Contribuții privind studiul micro și macroscopic al unor mortare de tencuială în contextul dezvoltării durabile. Teza de doctorat. Universitatea Tehnica din Cluj-Napoca, Facultatea de Constructii, Romania, 2012.
[9] Aciu Claudiu. Materiale ecologice în elemente structurale și nestructurale în construcția de locuințe. Teza de doctorat. Universitatea Tehnica din Cluj-Napoca, Facultatea de Constructii, Romania, 2009.
[10] Hussein G F et al 2020 *Alkali-activated mortars blended with glass bottle waste nano powder: Environmental benefit and sustainability* Journal of Cleaner Production **243**118636
[11] Obaid M K, Nasr M S, Ali I M, Shubbar A A F and Hashim K S 2021 *Performance of green mortar made from locally available waste tiles and silica fume* Journal of Engineering Science and Technology **16** 136 – 151
[12] Pintea A O, Manea D L, Aciu C, Jumate E and Fechete R 2020 *Plastering mortar with organic natural polymers studied by 1H NMR relaxometry* AIP Conf. Proc. 050004
[13] Jumate E, Moldovan D, Manea D L, Dan E D and Fechete R 2016 *The effect of cellulose ethers and limestone fillers in Portland cement-based mortars by 1H NMR relaxometry* Appl. Magn. Reson. **47** pp.1353–1373
[14] Saitis C, Manea L D, Plesa L M and Saitis L R *2023 Recycling Plaster Waste as a Substitute for Aggregates in Obtaining Plastering Mortars* Buildings **13(2)** 491
[15] Standard SR EN 998-1:2004. Specification of masonry mortars. Part 1: Mortar for plastering and smoothing
[16] Standard SR EN 998-1/AC-2006. Specification of masonry mortars. Part 1: Mortar for plastering and smoothing
[17] Standard SR EN 1015-2:2005. Methods of testing masonry mortars. Part 2: Global Sampling of Mortars and Preparation of Trial Mortars
[18] Tampieri A, Szabó M, Medina F and Gulyás H 2020 *A brief introduction to the basics of NMR spectroscopy and selected examples of its applications to materials characterization* Physical Sciences Reviews 20190086
[19] Siomin A-C PHD Thesis. Waste management and their capitalization in the field of building materials
[20] Tintisan M L PHD Thesis. Contributions on the implementation of cellulose waste mortars
[21] Standard SR EN 196. Test methods of cement. Part 1: Determination of mechanical strengths
[22] Manea D L, Netea A G and Aciu C 2019 *Materiale de construcții și chimie aplicată* UTPress Publishing House

Laboratory study regarding recipes and strengths of concrete produced with reactive powders and metallic fibers

S Scripcă[1*], M Bărbuță[1] and V Panaite[1]

[1] "Gheorghe Asachi" Technical University of Iasi, Faculty of Civil Engineering and Building Services, Iasi, Romania

* sabina.scripca@student.tuiasi.ro

Abstract. Because of its multiple advantages such as durability, flexibility in use, fire resistance and low production and maintenance costs, concrete is, by far, the most popular building material currently in use. However, cement – one of the base ingredients of concrete – requires large quantities of raw materials and fossil fuels in order to be produced. In fact, in terms of pollution, it is one of the most environment-damaging products of the construction industry. A relatively recent trend in the problem of mitigating cement-related pollution is the possibility to substitute a certain percentage of cement with reactive powders, which are considered to be waste products of other industries. Concurrently, another trend is to try and obtain better performance of plain concrete, by using so-called "disperse reinforcement". Thus, in the present article, a laboratory study is presented, regarding the obtainment and performance of concrete in which cement has been partially replaced with ultrafine silica and fly ash. Furthermore, in the case of the fly ash recipe, a percentage of metallic fibers was added, in order to study the potential performance gain with respect to plain concrete. The obtained results in compression were encouraging, in the sense that, for all recipes of concrete (plain concrete – used as reference sample versus concrete with fly ash, concrete with ultrafine silica, concrete with fly ash and fibers), the desired concrete strength class was attained. Even more, in the case of obtained results in tension by bending and splitting, the concrete with fly ash and fibers attained almost twice the tensile strength of the plain concrete. The obtained results suggest that concrete with reactive powders and/or metallic fibers can successfully substitute the plain concrete normally used in the construction industry.

1. Introduction

Much research has been devoted to the obtainment of so-called "special concrete". Some examples include: high-strength concrete, self-compacting concrete, polymeric concrete, and ecological concrete. These types of concrete managed to tackle several known issues of "regular concrete". Therefore, when one employs "special concrete", it is possible to obtain increases in mechanical strengths, improved behavior at high temperatures, better resistance to frost/thaw cycles, reduced capillarity and carbonation, high abrasion resistance, etc [1-3].

In the past few years, there is a greater emphasis in using waste material in concrete recipes, such as fly ash, ultrafine silica, furnace slag, crushed glass, various fibers, plastic materials, metakaolin etc. This is because the environmental problem becomes more pressing with each passing year. As a positive side effect, it was also observed that, in some cases, these materials also improved the properties of the obtained products. Therefore, there is a growing interest in the construction industry for the research and development of concrete recipes which incorporate waste materials. These waste materials can be used as partial replacements for cement or as additions to the concrete mixture, improving its properties, such as compressive strength, durability, permeability, and reducing its carbon footprint [4-7].

The aim of the present paper is to present several recipes and the subsequent testing results of regular concrete versus concrete with ultrafine silica, with fly ash and with disperse reinforcement (metallic fibers). A recipe which combined some of these materials was also made and tested – metallic fibers and fly ash.

2. Materials used

In this article, several types of materials were used to obtain a superior and environmentally friendly concrete, with characteristics comparable to the control concrete. The commercially available Portland cement CEM II/A-LL 42.5 R and three types of aggregates were used: sand with a grain size of 0-4 mm, crushed stone with a grain size of 4 - 8 mm and 8 - 16 mm. The water used was potable and complied with the SR EN 1008 standard.

To improve the properties of concrete, fly ash was added, in accordance with the SR EN 450 standard, and ultrafine silica, in accordance with the SR EN 13263 standard. Ultrafine silica is a fine powder with a high SiO_2 content, obtained from the technological process of ferrosilicon production. Due to the very fine particles, 100 times smaller than those of cement, ultrafine silica can be successfully used as a cement replacement and can form hydraulic hardening compounds that develop mechanically strong concrete. Fly ash, also a fine powder, is produced during the combustion of pulverized fuels and can vary in composition depending on the chemical composition of the fuel and the burning conditions. Both are waste materials that researchers are intensively investigating as cement replacements or concrete additives. To increase the tensile strength of the concrete, dispersed metallic fibers were used, in accordance with the SR EN 14889 standard. All used materials were carefully selected to ensure superior concrete quality. In addition, a Sika Plastiment BV-440 additive, a plasticizer/water reducer for concrete with universal applicability, was used.

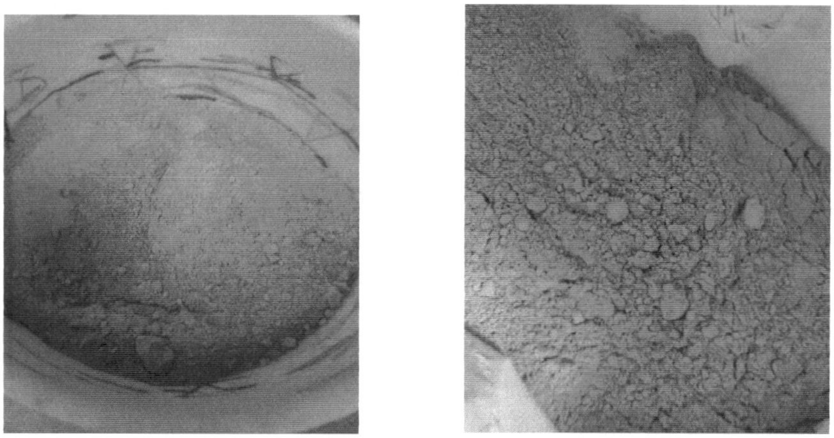

(a) (b)

Figure 1. Fly ash (a) and ultrafine silica (b).

The current state of research in the field indicates that ash and silica can be successfully used as cement substitutes in concrete, improving its properties and reducing its environmental impact. There are numerous studies that have investigated the effects of ash and silica on concrete performance, with positive results in terms of concrete strength, permeability and durability [8-11].

The availability of the two additives may vary depending on the geographical region and the availability of raw materials in that area. However, in general, fly ash may be more readily available in certain regions and cheaper because significant quantities may be found near coal-fired power plants. In contrast, ultrafine silica may be more difficult to obtain in certain regions because it is a specialized product obtained from complex industrial processes. Therefore, in general, ash could be considered easier to obtain than ultrafine silica.

Figure 2. Metallic fibers.

In recent years, numerous studies and research have confirmed the benefits of adding metal fibers to concrete. This field is quite advanced and well-developed. When added to concrete, metal fibers act as additional reinforcement, promoting a more even distribution of forces and preventing cracking. They are commonly used to enhance the mechanical properties of concrete, particularly its tensile strength. Metal fibers also improve the durability and performance of concrete under dynamic loading and impact [12-16].

3. Types of samples

The types of samples used are:

- concrete cubes with the side of 150 mm
- concrete prisms with the dimensions 100 x 100 x 550 mm

Aggregates with sizes of 0-16 mm were used for the production of concrete in this study and the following recipes were taken into account:

- Fly ash concrete with 10% substituted cement – figure 3

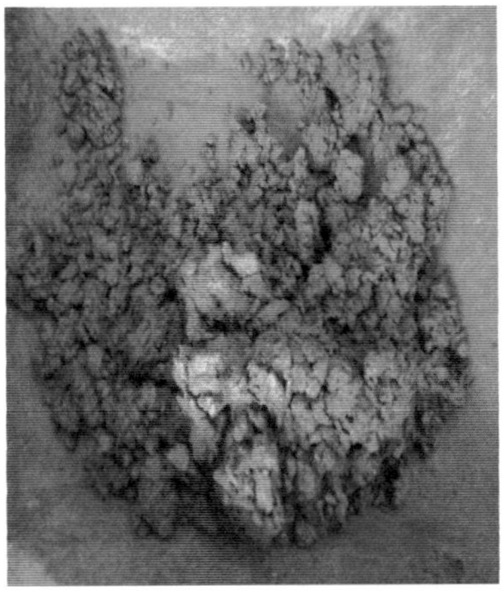

Figure 3. Photo of fly ash concrete.

- Ultrafine silica concrete with 10% substituted cement - figure 4

Figure 4. Photo of ultrafine silica concrete.

- Fly ash concrete with 10% substituted cement and added metallic fibres - figure 5

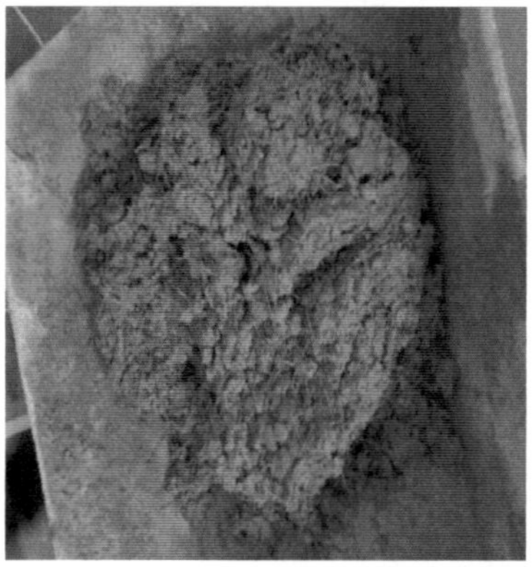

Figure 5. Photo of fly ash concrete and metallic fibers.

The control proportions of the concrete mixture are presented in tables 1 and 2.

Table 1. Concrete recipes for reference sample, C.

Aggregate 0-4 mm kg/m^3	Aggregate 4-8 mm kg/m^3	Aggregate 8-16 mm kg/m^3	Cement CEM II 42.5 R, kg/m^3	Water l/mc	Plasticizer additive
842	400	583	375	188	1% of the cement quantity

Table 2. Changes / additions to the reference recipe for FC, SC and MFC.

Ingredient kg/m3	concrete with fly ash FC	concrete with ultrafine silica SC	concrete with metallic fibers and fly ash MFC
CEM II	337.5	338.5	356.25
Fly ash	10% of the cement quantity: 37.5	——	5% of the cement quantity:18.75
Ultrafine silica	——	10% of the cement quantity: 37,5	——
Metallic fibers	——	——	3

4. Laboratory procedure

Following the pouring of the samples, they were kept (in accordance with SR EN 12390-2) for 24h at a temperature of 20 ± 2°C. After the first 24h they were stripped of their formwork and were kept submersed in water, at a temperature of 20 ± 2°C, until it was time to test them. The tests were made at 28 days and 3 samples were allocated to each test, according to current standards.

5. The determination of the apparent density of concrete after 28 days

The determination of the apparent density was made at 28 days. It was carried out on three different samples according to STAS 2414-91.

The formula for the density is:

$$\rho_a = \frac{m}{V_a} \tag{1}$$

Where:

- ρ_a – the apparent density
- m – the mass of the sample at natural humidity
- V_a – the apparent volume

The obtained values for densities are shown in the table below:

Table 3. Density determination at 28 days.

No.	Recipe name	Density [kg/m³]
1	C	2350
2	FC	2232
3	SC	2216
4	MFC	2335

6. The determination of the mechanical strengths

The mechanical strengths of concrete are usually established through short duration loading. A valid result is considered to be the average of three tests.

6.1. *The determination of the compressive strength*

The compressive strength is widely regarded as one of the most important properties of hardened concrete and it defines its class. The compression strength is considered a main quality criterion for concrete and it is determined according to SR EN 12390-3:2019.

The compressive strength is determined with the formula:

$$f_c = \frac{F_c}{A} \tag{2}$$

Where:

- F_c – the maximum force;
- A – the area of the surface perpendicular to the loading direction.

Figure 6. Compression testing and failure modes for the concrete samples.

The concrete cubes were tested in compression with a hydraulic press, acting perpendicular to the casting direction of the concrete, in accordance to the previously-mentioned standard.

The values obtained for the compression strength of the studied samples after 28 days can be consulted in the table below:

Table 4. Compression strength of concrete with 0-16 mm sort.

No.	Recipe name	(f_c) [N/mm^2]
1	C	31.43
2	FC	26.80
3	SC	26.99
4	MFC	27.52

6.2. The determination of the flexural strength

The flexural strength can be determined on simply supported prisms which are subjected to bending through the application of a concentrated force mid-span. The testing is done in accordance to SR EN 12390-5:2019. Failure will occur in the maximum bending moment section and the sample will split in two.

The flexural strength is determined with the following formula:

$$f_{ti} = \frac{M_p}{W_{pl}} = \frac{3}{2}\frac{F_{ti} \times l}{b \times h^2} \quad (3)$$

Where:

- F_{ti} – the applied force;
- l – the distance between the two supports;
- b, h – the two sides of the sample.

Figure 7. Flexural strength test and failure modes for the concrete samples

The prisms were tested for flexural strength with a hydraulic press, acting perpendicular to the concrete casting direction and the obtained value represents the average of three tested samples.

6.3. The determination of the splitting tensile strength

The splitting tensile strength can be established on concrete cubes, prisms and cylinders. For the current manuscript, only prism halves were used. The testing is done in accordance with SR EN 12390-6:2010.

In the case of tests being made on cubes, the following formula is used:

$$f_{td} = 0.64 \times \frac{F_{td}}{a^2} \tag{4}$$

Where:

- a – the size of the cube edge;
- F_{td} – the applied force.

The tests were made with a hydraulic press, acting perpendicular to the concrete casting direction and the obtained value represents the average of three tested samples.

Figure 8. Tension by splitting testing and failure mode for the concrete samples.

The values of the flexural strength and of the splitting tensile strength of the concrete, after the curing period of 28 days, can be consulted in the table below:

Table 5. Tensile strength of concrete.

No.	**Recipe name**	**Tension by bending** f_{ti} **[N/mm²]**	**Tension by splitting** f_{td} **[N/mm²]**
1	C	3.39	1.81
2	FC	2.55	1.48
3	SC	2.48	1.55
4	MFC	6.27	3.43

7. The interpretation of the results and associated conclusions

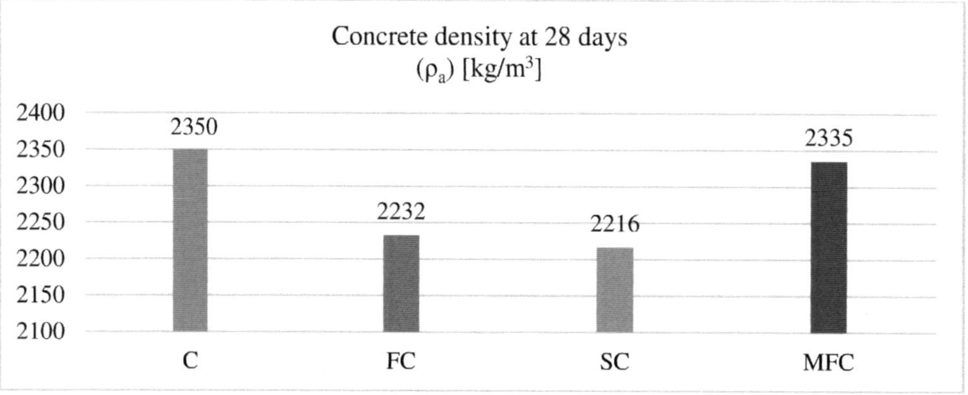

Figure 9. The density at 28 days for every type of concrete.

According to figure 9, the density of the studied types of concrete corresponds to that of concrete with normal volume mass (a density greater than 2000 kg/m^3 and lower than 2600 kg/m^3).

The smallest density is 2216 kg/m^3. This density was obtained for the SC recipe, in which 10% of the cement was substituted with ultrafine silica. The overall densities are situated between 2216 kg/m^3 and 2235 kg/m^3, therefore a lower density than the one registered for the reference sample can be noted.

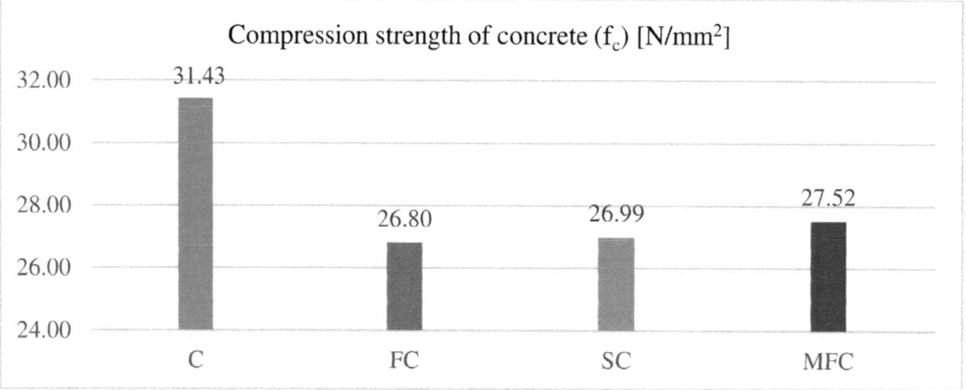

Figure 10. The compression strength at 28 days for every type of concrete.

Figure 10 shows that the lowest compression strength, of 26.08 N/mm^2, pertains to the FC concrete recipe (where 10% of the concrete quantity was substituted with fly ash) and the highest compression strength, of 27.52 N/mm^2, pertains to the MFC concrete recipe (where 5% of the concrete quantity was substituted with fly ash and metallic fibers were added).

Overall, the compression strength has a decreasing tendency, especially for FC and SC recipes. For the MFC recipe, the metallic fibers allowed for some improvement in the tensile behavior.

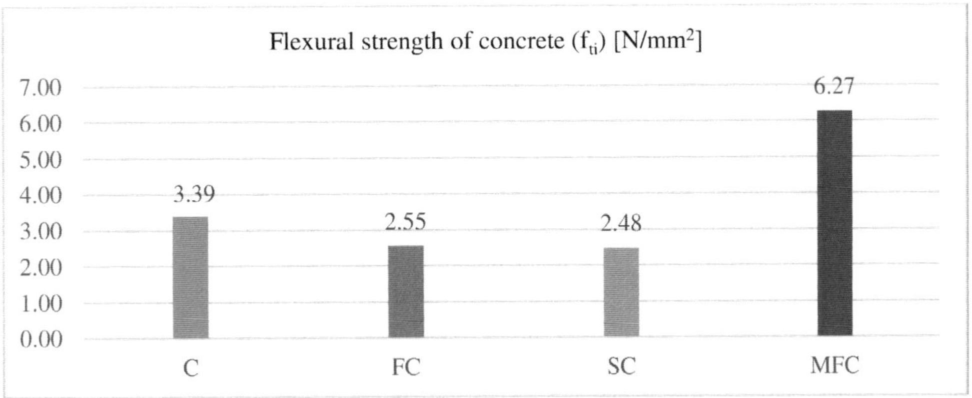

Figure 11. The tension by bending strength at 28 days for every type of concrete.

Figure 11 shows that the lowest flexural strength, of 2.48 N/mm2, pertains to the SC concrete recipe (where 10% of the concrete quantity was substituted with ultrafine silica) and the highest flexural strength, of 6.27 N/mm^2, pertains to the MFC concrete recipe (where 5% of the concrete quantity was substituted with fly ash and metallic fibers were added).

Overall, the flexural strength has a decreasing tendency, especially for FC and SC recipes, where the strengths were, on average, approximately 25% lower than the ones obtained for the reference samples. For the MFC recipe, the metallic fibers allowed for a large improvement in the tensile behavior, with the strength increasing by approximately 84%.

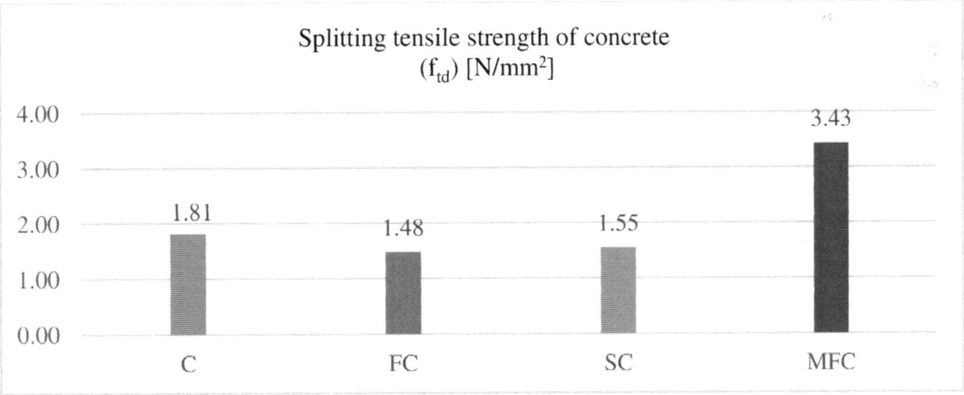

Figure 12. The splitting tensile strength at 28 days for every type of concrete.

From figure 12 it results that the lowest splitting tensile strength, of 1.48 N/mm^2, pertains to the FC concrete recipe (where 10% of the concrete quantity was substituted with fly ash) and the highest tensile by bending strength, of 3.43 N/mm^2, pertains to the MFC concrete recipe (where 5% of the concrete quantity was substituted with fly ash and metallic fibers were added).

Overall, the splitting tensile strength also has a decreasing tendency, especially for FC and SC recipes, where the strengths were, on average, approximately 16% lower than the ones obtained for the reference samples. For the MFC recipe, the metallic fibers allowed for a large improvement in the tensile behavior, with the strength increasing by approximately 89%.

8. Final conclusions

During the preparation of the recipes, it was observed that the addition of fly ash led to a greater absorption of water, which determined an increase in the viscosity of the paste. This led to a decreased workability when compared to other recipes which did not include fly ash.

Following the tests, lower strengths were obtained – but still close the ones registered on the reference samples, because of the reduced quantity of cement.

The addition of metallic fibers led to the obtainment of positive results, with large increases in the tensile strengths, both in bending and in splitting. This proves that the metallic fibers do have the capacity to take over the tensile efforts induced through the testing loads.

The best results were obtained by the MFC recipe, where there was only a 12% reduction in compression performance compared to the reference samples. At the same time, it was observed that metallic fibers do not increase the compressive strength considerably when comparing FC and MFC, with an increase of only 2.68% when adding fibers.

The main advantages of using fly ash and ultrafine silica as substitutes for cement are represented by the technological improvements in the concrete, as it allows for the obtainment of reduced permeability, better resistance to frost-thaw cycles and ecological benefits.

The main advantages of using metallic fibers are represented by the increases in mechanical strengths and better control of the cracking processes – the fibers allow for the delay of the cracking process, which equates to a higher ductility, because the efforts in the tensed sectors are mainly constant.

Finally, it can be stated that thermal/silica ash and/or metal fibers concrete could be used in a variety of construction applications such as road and bridge construction, paving, masonry and structural elements. Furthermore, they could be a viable alternative for rehabilitation works, where the use of traditional concrete might not be feasible or sustainable. The development of sustainable and durable building materials is essential in today's construction industry, and the use of fly ash and silica as cement substitutes, combined with steel fibers, offers a promising solution for achieving these goals. However, it is important to carry out more studies and tests to develop and improve the properties of these concretes and to extend their range of applications in construction. Further, research is also needed to investigate the long-term durability and resistance to environmental factors of this type of concrete, as well as the potential for its use.

Acknowledgments

This paper was financially supported by the Project "Network of excellence in applied research and innovation for doctoral and postdoctoral programs / InoHubDoc", project co-funded by the European Social Fund financing agreement no. POCU/993/6/13/153437.

Reference list

[1] Peerzada D and Mohan Ganesh G 2020 Behaviour of self-compacting concrete using different mineral powders additions in ternary blends Romanian Journal of Materials 50 232-239

[2] Shukla A, Gupta N, Gupta A, Goel R and Kumar S 2020 Study on the behaviour of green concrete by the use of industrial waste material *IOP Conf. Ser.: Mater. Sci. Eng.* **804** 012036

[3] Shanmugapriya T and Uma R N 2013 Experimental investigation on silica fume as partial replacement of cement in high performance concrete *The International Journal of Engineeringand Science* **2** 40-45

[4] Annune J E, Mamodu O, Caesar S M and Ahmed Y A 2022 Investigating the chemical durability of concrete produced with pozzolanic additives *Journal of Emerging Technologies and Innovative Research* **9** 354-360

[5] Jalilifar H and Afshar R 2016 Experimental effect of using silica fume and fly ash on mechanical properties of recycled concretes *International Journal of Engineering and Applied Sciences* **3** 67-71
[6] Liew Y M 2011 Investigating the possibility of utilization of kaolin and the potential of metakaolin to produce green cement for construction purposes–A review *Australian Journal of Basic and Applied Sciences* **5** 441-449
[7] Brahami Y, Saeidi A, Fiset M and Ba K 2022 The Effects of the Type and Quantity of Recycled Materials on Physical and Mechanical Properties of Concrete and Mortar: A Review *Sustainability* **14** 14752
[8] Zhao S and Zhang Q 2019 Effect of Silica Fume in Concrete on Mechanical Properties and Dynamic Behaviors under Impact Loading *Materials Basel* **12** 3263
[9] Nayak D K, Abhilash P P, Singh R, Kumar R and Kumar V 2022 Fly ash for sustainable construction: A review of fly ash concrete and its beneficial use case studies *Cleaner Materials* **6** 100143
[10] Farmani F, Khadiv-Parsi P, Ramezanianpour A A, Bonakdarpour B and Yazdian F 2022 Dual eco-friendly application of silica fume and scoria in cement-based materials through the the enhancement of microbially-induced carbonate precipitation *Case Studies in Construction Materials*, **17** 01481
[11] Golewski G L 2023 Combined Effect of Coal Fly Ash (CFA) and Nanosilica (nS) on the Strength Parameters and Microstructural Properties of Eco-Friendly *Concrete Energies* **16** 452
[12] Pan Z, Zeng X and Hu Z 2019 Effect of fly ash and silica fume on mechanical properties of high performance FRCC *IOP Conf. Series: Materials Science and Engineering* **562** 012048
[13] Facconi L and Minelli F 2016 Verification of structural elements made of FRC only: A critical discussion and proposal of a novel analytical method Engineering *Structures* **131** 530-541
[14] Moussaceb K Use of metallic fibers in concretes 2014 *MATEC Web of Conferences* **11** 01027
[15] Anwar A G 2016 A Study on Mechanical Properties of Metallic Fiber Reinforced Concrete *International Journal of Engineering Technology Science and Research* **3** 12-24
[16] Osman G 2011 Workability and Mechanical Performance of Steel Fiber-Reinforced Self - Compacting Concrete with Fly Ash *Composite Interfaces* **18** 169–184

IOP Conf. Series: Materials Science and Engineering 1283 (2023) 012011 doi:10.1088/1757-899X/1283/1/012011

Steel hall constructive solution beam with a full web, optimization of sections and joints

F Tepes-Onea[1*] and N Janczak[1]

[1] Ovidius University of Constanta, Faculty of Civil Engineering, Romania

* tflorin@univ-ovidius.ro

Abstract. In the present paper, the design of a steel warehouse structure was investigated. After establishing the base sections for the beams and columns through the design for ultimate resistance and buckling strength, an optimization procedure was applied to the beam-to-column connection. Three types of connections were studied: a bolted connection with end plate, a welded connection and a bolted connection with web and flange angles. The detailed analysis of the components of the connections was performed with the help of the software "IdeaStatica". The advantages and disadvantages of the three joint solutions are presented. The results were compared and the best solution was proposed in view of costs, structural performance, inspection difficulties, joint stiffness, and the installation and safety process.

1. Introduction

The present paper aims at optimizing the beam-to-column connection by analyzing the stress levels and constructive advantages of three structural joint solutions. The optimization of the column and beam sections was also considered, as the M-N interaction relationship reached the limit value of 1.00.

The steel structure consists of portal frames with a span of 16.7 m and a frame spacing of 5 m. The portal frame features a nominally pinned base and a moment resisting eaves and apex connections.

The moment-resisting beam-to-column connection is designed to avoid reaching plastic behavior; if such a stress level is reached, the plastic limit should be first reached and limited to the beam or column sections. For this purpose, several design approaches may be applied.

There are many technical solutions for moment resisting connections. In this paper, three types of connections [1] between I-shaped beams and H-shaped columns [2-4] were analyzed:

- bolted end-plate connection;
- welded connection;
- bolted connection with web and flange angles.

For elements subjected to both bending and compression, interaction relationships were used that must meet the condition ≤ 1.

The interaction formulas are based on the modelling of simply supported elements with fork supports and with or without continuous lateral constraints, which are then subject to compressive forces, bending moments and/or transverse loads.

The effects of slenderness and moment distribution cannot be easily derived on a purely physical reasoning and are accounted for by means of additional terms – k_{yy}, k_{zz}, k_{yz}, k_{zy} (section classes 1 & 2) [5,6].

1.1. Types of moment-resisting joints

1.1.1. Bolted end-plate connection. The main characteristics of this type of joint (ultimate resistance and stiffness) depend on its components. The main factors that influence the behavior and characteristics of this type of joint are: the type of end-plate used (standard, unstiffened extended or stiffened extended); the thickness of the end-plate "t"; the diameter of the bolts; the presence of a column stiffening plate (subjected to compression); the shear stiffening of the column's panel [6].

1.1.2. Welded connection. The welded connection provides a high degree of strength and stiffness. By providing a weld with larger resistance than that of the base material, the yielding process will start in the beam element and never reach the joint itself. The stiffness of the connection is dictated by the flexibility of the column's panel (implicitly, by its thickness); improvements in this regard could be obtained by the insertion of stiffening plates on the column's web panel [6].

1.1.3. Bolted connection with web and flange angles. This type of connection uses angles to transfer the stresses from one element to the other: tension and compression stresses are transferred through the top and bottom angles, attached to the beam flanges, and shear stresses are transferred through the clip angles attached to the beam web. The main problems of these types of joints are the potential sliding of the bolts in the holes of the flanges and of the angles and the behavior of the angle profiles subjected to bending [3].

1.2. Verification relationship for M-N interaction.

The beam-to-column connection is subjected to a combination of forces and unidirectional or bidirectional bending moments. The bending moments are transferred from the beam, but may be generated by transverse forces acting between the ends of the elements too. This connection ensures the axial compression of the column, by transferring the bending moments and shear forces from the beam. It is the most common type of joint.

The behavior of a beam-to-column connection is usually treated as an isolated structural member, subjected to bending moment and axial stress. The evaluation of this interaction involves the use of k_{yy} k_{yz}, k_{yz} and k_{zz} factors. K factors take into account the post-elastic behavior in the bending moment – axial force interaction [5].

2. Structural analysis

2.1. Choosing the structural concept

The structure considered for this study is a portal frame structure with a single span of 16.7m, a distance between frames of 5.0 m and a column height of 7.0 m. Standard IPE330 profiles were used for the beams and HEB300 profiles for the columns. The structure is made out of S355 grade steel. The main loads are presented in table 1.

Table 1. Fundamental load combination.

Element	Intensity basic standard [kN/ m^2]	Length / Surface [m^2]	Loading norm [kN/m]	The action coeff.	Loading calculation [kN/m]
Sandwich panel covering	0.32	5	1.6	1.35	2.16
IPE 330 beam	0.491	1	0.491	1.35	0.66
Snow	1.20	5	6	1.50	9.00
				Σ =	**11.82 kN/m**

Table 2. Internal forces for the beam with constant section.

Element	Axial force N_{Ed} [kN]	Shear force V_{Ed} [kN]	Bending moment M_{Ed} [kN·m]
Beam with constant section	46.9	99.2	269.1

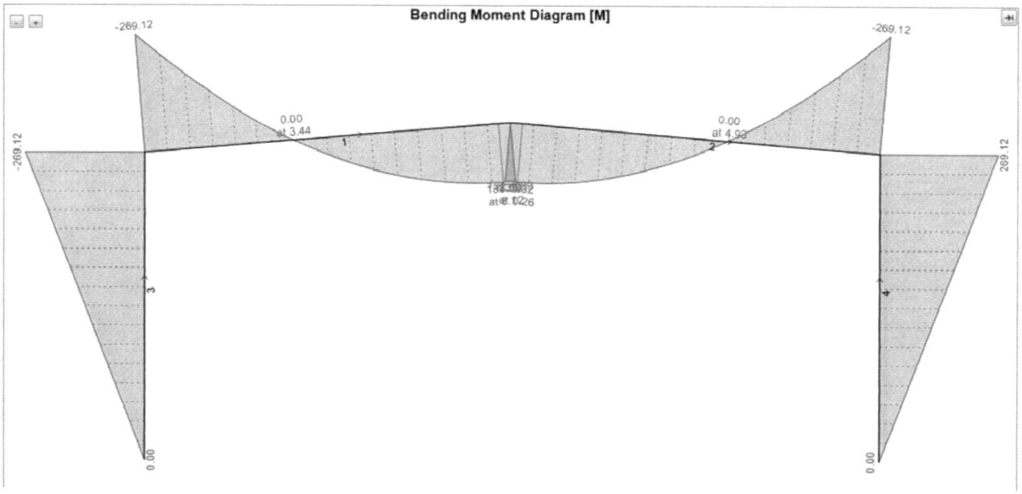

Figure 1. Bending moment diagram (selected values in table 2).

2.2. Resistance check, buckling and limit displacement

The design values of the internal forces (N_{Ed}, M_{Ed}, V_{Ed}) for the IPE300 beams are presented in table 2 and figure 1 and should satisfy [1]:

$$\frac{V_{Ed}}{V_{c,Rd}} = \frac{99.2}{631.3} = 0.157 < 0.5 \Rightarrow \text{the section is verified} \tag{1}$$

$$\frac{N_{Ed}}{N_{c,Rd}} = \frac{46.9}{2222.3} = 0.021 < 0.1 \Longrightarrow \text{the section is verified} \tag{2}$$

$$\frac{M_{Ed}}{M_{c,Rd}} = \frac{269.1}{285.53} = 0.94 < 1 \Longrightarrow \text{check} \tag{3}$$

For the buckling resistance check, members subjected to combined bending and axial compression should satisfy: [5]:

$$\frac{N_{Ed}}{\frac{\chi_y \cdot N_{Rk}}{\gamma_{M1}}} + k_{yy} \cdot \frac{M_{y,Ed}}{\frac{\chi_{LT} \cdot M_{y,Rk}}{\gamma_{M1}}} = \frac{46.9 \cdot 10^3}{\frac{0.63 \cdot 2222.3 \cdot 10^3}{1}} + 1.02 \cdot \frac{269.1 \cdot 10^6}{\frac{0.96 \cdot 285.53 \cdot 10^6}{1}} = 1 \leq 1 \tag{4}$$

$$\frac{N_{Ed}}{\frac{\chi_z \cdot N_{Rk}}{\gamma_{M1}}} + k_{zy} \cdot \frac{M_{y,Ed}}{\frac{\chi_{LT} \cdot M_{y,Rk}}{\gamma_{M1}}} = \frac{46.9 \cdot 10^3}{\frac{0.71 \cdot 2222.3 \cdot 10^3}{1}} + 0.528 \cdot \frac{269.1 \cdot 10^6}{\frac{0.96 \cdot 285.53 \cdot 10^6}{1}} = 0.54 \leq 1 \tag{5}$$

Both relationships are satisfied.

The verification of Service Limit Status (SLS) is presented below [5]:

$\delta = 48\ mm$ – maximum vertical displacement

$$\delta < \frac{L}{250} = \frac{16700}{250} = 66.80\ mm \tag{6}$$

$48 < 66.8 \Longrightarrow$ the section is verified

2.3. Optimization of nodes with the software "IDEA StatiCa"

To verify and optimize the beam-to-column connection, IDEA StatiCa was used. The shear force and bending moment from figure 1 were applied on the connection. The types and dimensions of the specific components of the connections are detailed for each case separately.

2.3.1. Bolted end-plate connection. In order to achieve better results (table 3), stiffening plates were added. For the end plate, a thickness of 25 mm was chosen. The thickness of the weld is 11mm. M24 bolts and quality group 8.8 were chosen.

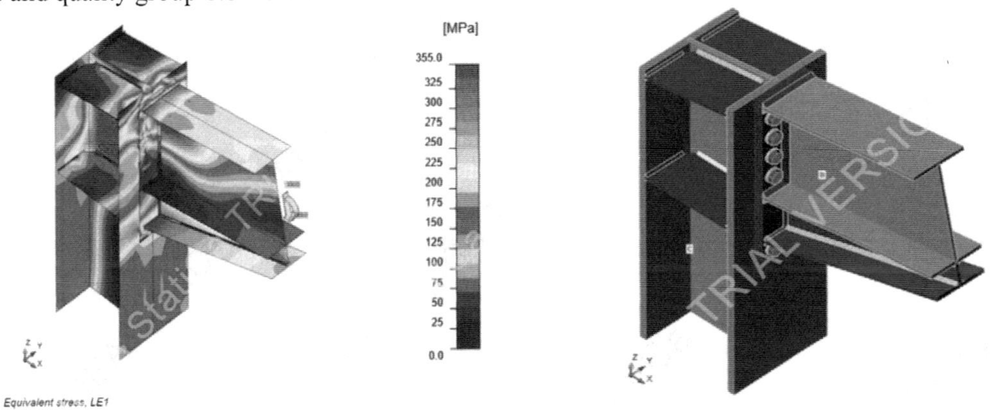

Figure 2. Equivalent stress and the 3D model of bolted end-plate connection.

Table 3. Summary for joint with end plate.

Name	Value	Status
Analysis	100%	Ok
Plates	0.3 < 5%	Ok
Bolts	90.4 < 100%	Ok
Welds	99.7 < 100%	Ok

2.3.2. Welded connection. The butt weld with a thickness is 15mm on both sides was applied. In order to obtain better results, 18 mm thick stiffeners were added (figure 3 and table 4).

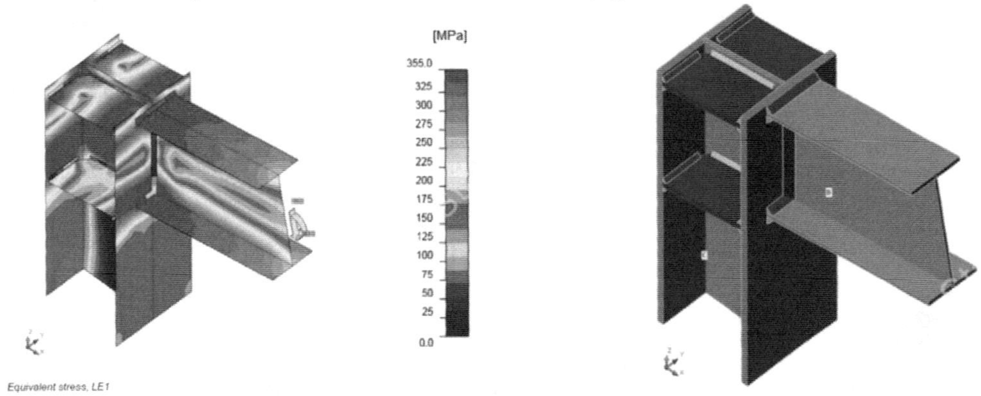

Figure 3. Verification of equivalent stress. 3D model of welded joint.

Table 4. Summary for joint with welded.

Name	Value	Status
Analysis	100%	Ok
Plates	2.6 < 5%	Ok
Welds	99.3 < 100%	Ok

2.3.3. Bolted connection with web and flange angles. After many optimization attempts, the final test (table 5, figure 4) was performed with a continuous column, two L150x100x12 angles connected with two M16 8.8 bolts on the lower and upper flange and two clip angles of the same size with M16 8.8 four bolts fixed on both sides of the beam web.

Table 5. Summary.

Name	Value	Status
Analysis	70.3% ≤ 100%	Ok
Plates	397.6 % > 5 %	no
Welds	653.4 % > 100 %	no

Even after all of these attempts, the check for this type of connection could not be completed.

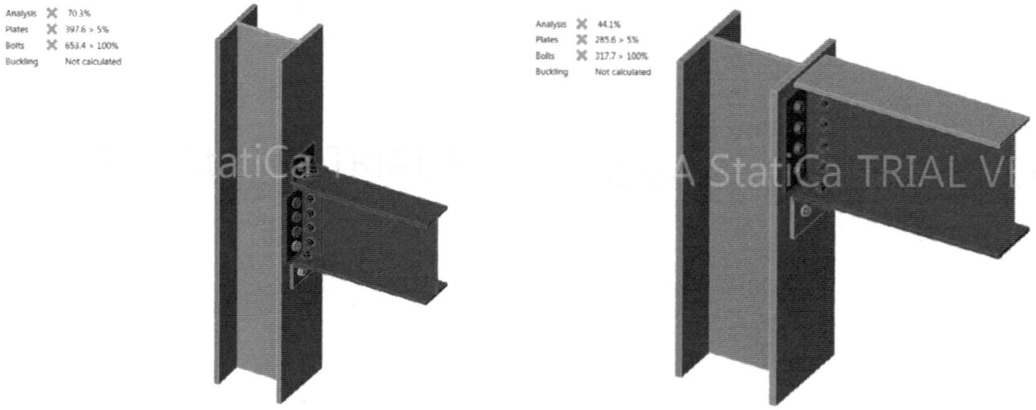

Figure 4. Two of the models used for the bolted connection with web and flange angles.

A comparison of the final results is presented in table 6 for the three types of joints. One of the conclusions that can be drawn is that the only possibilities for a moment-resisting connection are the welded joint and bolted joint with end plate. The optimization of the sections and the connection itself was successful, as the utilization factor is close to 1.00.

Table 6. Results obtained from IdeaStatica.

Objective	Welded joint	Joint model with end plate	Joint model with angle plates	Condition
Analyse	100	100	100	> 100%
End plate	2.6	0.3	588.5	<5%
Welds	99.3	90.4	N/A	<100%
Bolts	N/A	99.7	511.5	<100%

3. Conclusions

The optimization of the sections was carried out through an analytical, iterative calculation, until the M-N interaction checks reached a value close to 1.00. Thus, the sections IPE330 for the beam and HEB300 for the column were obtained. This optimization was achieved by determining, for each of the chosen section, the factors k_{yy} k_{yz}, k_{yz} and k_{zz} aimed at taking into account the effects of post-elastic behavior in the interaction between bending moment and axial force.

In order to reach a suitable bolted end-plate connection, a batten and stiffener plates were added, the end-plate thickness and the size of the bolts were increased and better quality bolts were chosen. In the case of the welding joint, the following changes were made to the thickness of the weld and the distance between the end of the column and the beam and stiffening plates were added.

For the same beam and column sections that are subjected to the same loads, the bolted end-plate connection could not have been successfully designed without the addition of a batten.

The maximum stress level for both joints is the same, with a yield at 355 MPa. In the case of the welded joint, the yield limit is reached and extends over the entire node and over the beam (figure 3).

For the bolted beam-to-column connection with web and flange angles, the check could not be finalised, even after multiple attempts.

In conclusion, bolt joints have the advantage of cost and simplicity, but there is a loss of structural performance. On the other hand, welded joints are more resistant but more expensive and harder to handle.

Acknowledgments
Rapid Earthquake Damage Assessment Consortium-REDACt [BSB 966].
Contract No: MLPDA 88712/26.06.2020.

References

[1] Dubina D, Florea D 2010 *The global structural design of the stell structures. Recommendations, comments and examples of application according to SR EN 1993-1-1 and SR EN 1998-1*
[2] Uang CM, Bruneau M, P. Eng, *Seismic Design of Steel Structures, Chapter 9*
[3] Maquoi R, Boissonnade N, Muzeau J P, Jaspart J P and Villette M 2001 The interaction formulae for beam-columns: A new step of a yet long story *Proc. - Annu. Tech. Sess. Struct. Stab. Res. Counc.* 63–88
[4] Yang B and Tan K H 2012 Numerical analyses of steel beam-column joints subjected to catenary action *J. Constr. Steel Res.* **70** 1–11
[5] Dubina D, Ungureanu V, Zaharia R and Dogariu A 2010 *Verificarea la stabilitate a elementelor din otel în conformitate cu SR EN 1993-1.1 Recomandări de calcul, comentarii şi exemple de aplicare (Checking the stability of a steel element in accordance with SR EN 1993-1.1, Recommendations for calculating comm* (Timisoara)
[6] Dubina D, Grecea D 2009 *Calculation and projection of joints structural steel in conformity SR-EN 1993-1-8*
[7] Kober H, *The behavior under seismic actions of steel structures equipped with different bracing types, AICR*

Identification method of unbalanced fault characteristics of horizontal shaft of rotary table of drilling rig

Z Wang[1*], C Q Lu[1], J Ma[1] and S Y Du[1]

[1] School of Mechanical Engineering, Shenyang Jianzhu University, China

* juven1126@163.com

Abstract. The horizontal shaft of rotary table is an important rotating part of drilling rig. The vibration caused by the mass imbalance of the horizontal shaft will seriously affect the normal operation of drilling rig and lead to faults. Therefore, taking the horizontal shaft rotor system of rotary table of the drilling rig as the research object, the unbalanced vibration characteristics of the rotor are extracted by the improved all-phase fast Fourier transform method. Through the cross-correlation operation of the signal, the data of the discrete signal is weighted and superimposed, the vibration amplitude, time delay of the phase and the spectrum analysis can be obtained. The experimental results show that the accuracy and stability of the vibration amplitude and phase extracted by the improved all-phase fast Fourier transform method are better, and the accuracy of the vibration phase extraction can reach 98%, which verifies the effectiveness of the proposed method.

1. Introduction

A drilling rig is a large construction machine often used for exploration or mineral resource development. The rotary table in its rotating system can transmit the engine power to the downhole drilling tool through the horizontal shaft, so as to meet other downhole requirements such as rock breaking. But in the actual operation process, due to some factors such as materials, faults or assembly, the horizontal shaft of the drilling rig may have mass unbalance, which will cause vibration. The vibration of the drilling rig rotor system will greatly reduce the working accuracy of the drilling rig and even affect the construction safety.

In order to reduce the vibration of the horizontal shaft, it is necessary to perform mass compensation on the rotor system of the horizontal shaft, that is, dynamic balance. The premise of dynamic balance is to determine the unbalanced vibration characteristics of the horizontal shaft. Therefore, the vibration characteristics of the horizontal shaft need to be extracted first. The extraction and analysis of vibration characteristics of rotating machinery has always been a research hotspot in the field of machinery, and some achievements have been achieved.

Zhou proposed a new method based on wavelet packet multi-band spectral subtraction to enhance the characteristics of drilling processing vibration signals [1].

Xu proposed a method based on improved adaptive redundant lifting wavelet to improve the extraction accuracy of vibration signals [2].

Xie proposed an improved wavelet threshold denoising method to process vibration signals [3].

Wang Zhan et al. compared the vibration responses of the spindle system under experimental and simulated conditions, verified the accuracy of amplitude and phase extraction by cross-correlation algorithm [4].

To sum up, on the basis of the predecessors, this paper takes the improved all-phase fast Fourier transform method to extract the characteristics of the vibration signal of the rotating machinery, and obtains a more accurate rotor vibration and the unbalance of the equivalent position, which provides the basis for dynamic balance.

2. Unbalanced vibration feature extraction method based on improved All-phase Fast Fourier transform

In this method, the all-phase fast Fourier transform is improved in two steps. First, the amplitude of the fundamental frequency signal obtained by the cross-correlation operation is as follows:

$$e = \frac{2\sqrt{R_{yg}(0)^2 + R_{yh}(0)^2}}{T} \tag{1}$$

Collect 2N-1 sampling points from the time interval of the fundamental frequency signal $y(t)$, and then list the N N-dimensional vectors of the time series x_p according to formula (2);

$$\begin{cases} x_0 = [x(0), x(1), \ldots, x(N-1)]^T \\ x_1 = [x(-1), x(0), \ldots, x(N-2)]^T \\ \quad\ldots \\ x_p = [x(-p), x(-p+1), \ldots, x(N-1-p)]^T \\ \quad\ldots \\ x_{N-1} = [x(-N+1), x(-N+2), \ldots, x(0)]^T \end{cases} \tag{2}$$

where, x_p represents the N N-dimensional vectors of the PTH time series in the fundamental frequency signal $y(t)$, $x_{(-p)}$ represents the PTH sampling point with equal time interval, p satisfies p=0,1,2,…,N-1;

Move each $x(0)$ in the time series in formula (2) from x_0 to x_{N-1} to the first place, and obtain N N-dimensional vectors as shown in formula (3):

$$\begin{cases} x'_0 = [x(0), x(1), \ldots, x(N-1)]^T \\ x'_1 = [x(-1), x(0), \ldots, x(N-2)]^T \\ \quad\ldots \\ x'_p = [x(-p), x(-p+1), \ldots, x(N-1-p)]^T \\ \quad\ldots \\ x'_{N-1} = [x(-N+1), x(-N+2), \ldots, x(0)]^T \end{cases} \tag{3}$$

After $x(0)$ is aligned, add the N vectors in formula (3) to all the collection points in one dimension and take the average value, then the full-phase data vectors x_{ap} can be obtained:

$$x_{ap} = \frac{1}{N}\begin{bmatrix} Nx(0) \\ (N-1)x(1) + x(-N+1) \\ \ldots \\ x(N-1) + (N-1)x(-1) \end{bmatrix} \tag{4}$$

The discrete Fourier transform of x'_p in formula (3) is defined as $X'_p(k)$, and the discrete Fourier transform of x_p in formula (1) is defined as $X_p(k)$. According to the shift nature of DFT, the relation of $X'_p(k)$, and $X_p(k)$, can be obtained as:

$$X'_p(k) = X_p(k)e^{j\frac{2\pi}{N}vp} \tag{5}$$

where, v represents the number of harmonics, v satisfies $v=1,2,\ldots,N-1$ and j represents the complex number.

Take the sum of $X'_p(k)$ and average them, the output of the all-phase FFT is shown in formula (6), where φ_0 is the phase value of the unbalanced vibration signal processed by improved all-phase FFT;

$$X_{ap}(k)=\frac{1}{N}\sum_{p=0}^{N-1}X'_p(k)=\frac{1}{N}\sum_{p=0}^{N-1}X_p(k)e^{j\frac{2\pi}{N}vp}=\frac{e^{j\varphi_0}}{N^2}\cdot\frac{\sin^2[\pi(\beta-k)]}{\sin^2[\pi(\beta-k)/N]} \quad (6)$$

where, $\beta-k$ represents the frequency offset.

3. Simulation analysis of unbalance vibration characteristics identification of motorized spindle

In order to verify the advantages of the improved all-phase fast Fourier transform method in extracting the unbalanced fault features of vibration signals, this paper takes the common rotating mechanical electric spindle as the simulation and experimental object, respectively adopts the Fourier transform method, the mutual power method, The cross-correlation method and the improved all-phase fast Fourier transform method are used to extract the vibration characteristics in different states and compare them.

When the unbalanced mass phase is 45°, the rotor rotates clockwise, 60g·mm unbalanced mass is added to the spindle system, the static eccentricity is 0.0 mm, the spindle speed is 15000 r/min.

After the spindle runs smoothly, collect the unbalanced vibration signal in the X direction. Figure 1 shows the simulation results and signal fitting diagram of the unbalanced fault model of the motorized spindle. The abscissa represents time, the ordinate represents amplitude, the black point is the image of the original function, and the red curve is the image of the fitted function.

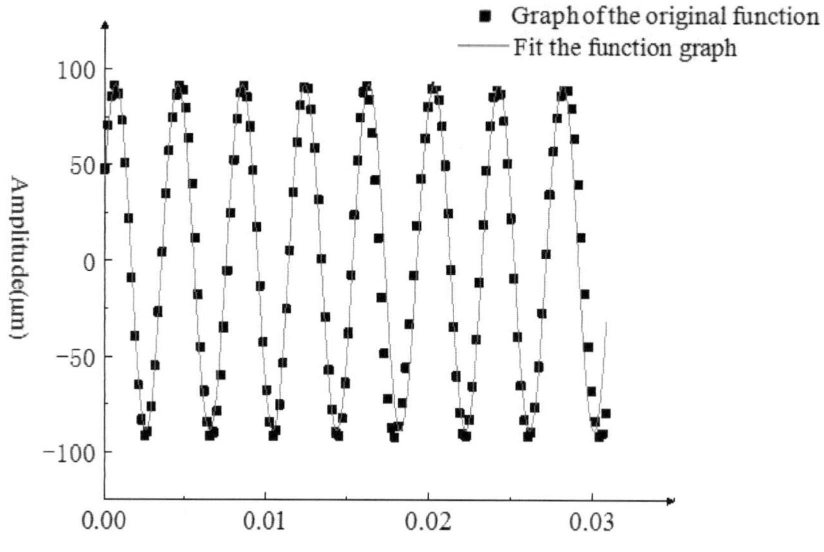

Figure 1. Fitting image of simulation data.

Add a Gaussian white noise signal with a signal-to-noise ratio of 1100 and a high-frequency signal commonly used in the environment into the fitting equation to obtain a new signal simulation equation:

$$\begin{aligned} y=&1.74514\times10^{-6}+9.04776\times10^{-5}\times\sin\left(1602.04t+0.24747\pi\right)\\ &+1\times10^{-5}\times\sin\left(16*1602.04t+0.24747\pi\right)\\ &+2\times10^{-5}\times\sin\left(32*1602.04t+0.24747\pi\right)+s(t) \end{aligned} \quad (7)$$

Table 1. Comparison of feature extraction methods.

	Fourier transform	Mutual power	Sinusoidal approximation	Improved all phase FFT
The average amplitude	93 μm	91 μm	96 μm	90 μm
The average phase	42.4°	43.0°	46.5°	45.0°

The simulation analysis shows that the vibration amplitude extracted by the improved all-phase fast Fourier transform method is most similar to the simulated signal amplitude among the four methods, and the interference information is removed. Moreover, the extracted phase information inherits the advantages of the all-phase fast Fourier transform, which is superior to the other three methods.

4. High speed spindle vibration signal extraction experiment

The experiment was verified by the electric spindle experimental platform, which is shown in figure 2. The motor spindles are two custom HT-170-20000-11 eccentric motor spindles with air gap eccentricity of 0.1 mm and 0.15 mm. The OFV-505/5000 Xtra laser vibrometers from Polytec GmbH are used to measure unbalanced vibrations of electrospindles during operation.

Due to the limitation of the actual experimental conditions, the vibration of the rotor is reflected by monitoring the exposed shaft at the end of the electrospindle, and the experimental data is derived by using the data display instrument.

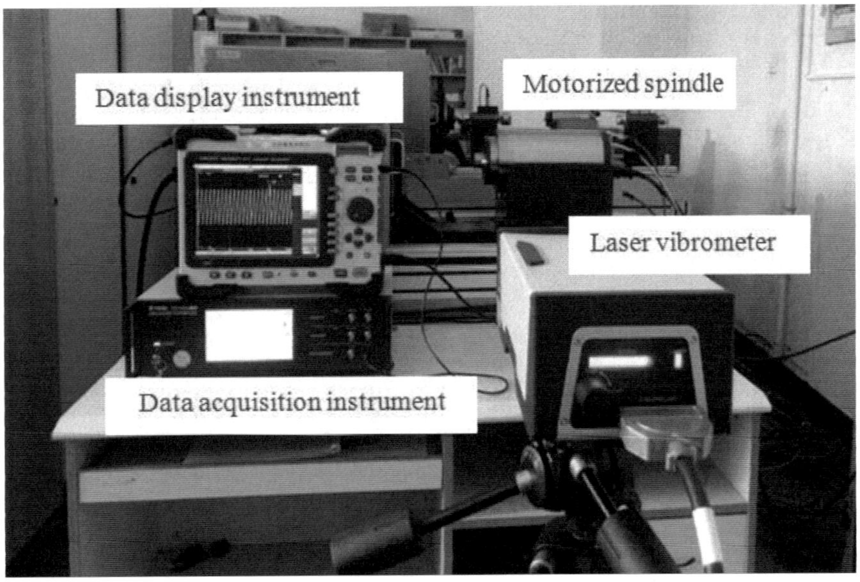

Figure 2. Motorized spindle experimental platform.

In the experiment, taking the improved all-phase fast Fourier transform method to extract the vibration signal, and it was compared with the other three vibration signal extraction methods, the comparative analysis chart of amplitude and phase was obtained, as shown below.

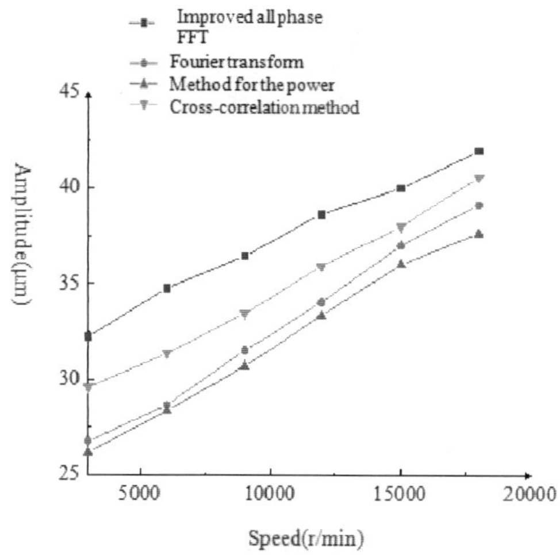

Figure 3. Amplitude of 30g·mm eccentric mass.

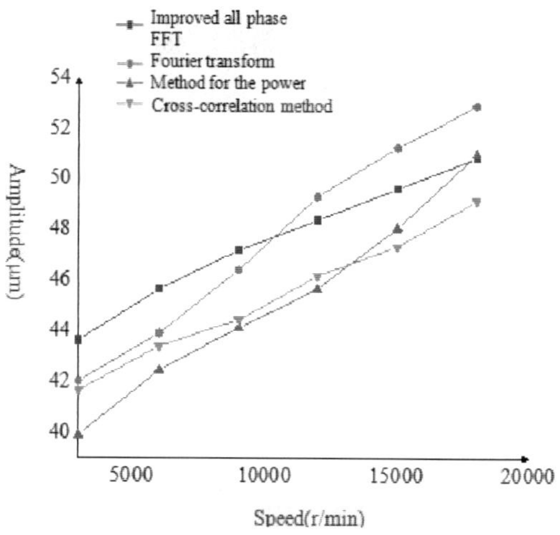

Figure 4. Amplitude of 40g·mm eccentric mass.

Figures 3 and 4 show a comparison of the amplitudes. It can be seen that with the increase of rotational speed, the amplitude of the vibration signals extracted by the four vibration signal extraction methods increases, and all fluctuate within a certain range.

As the unbalanced mass added to the high-speed spindle increases, the amplitude of the vibration signal at the same rotational speed also increases, and the amplitude is almost the same. It can be seen that both the improved all-phase fast Fourier transform method and other methods can accurately extract the amplitude of the main shaft vibration.

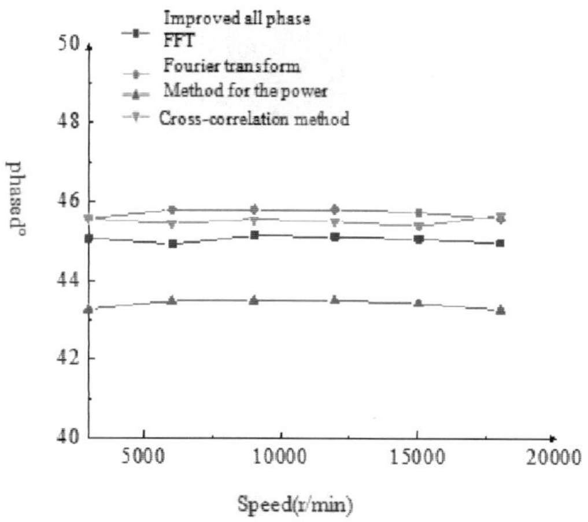

Figure 5. Phase of 30g·mm eccentric mass.

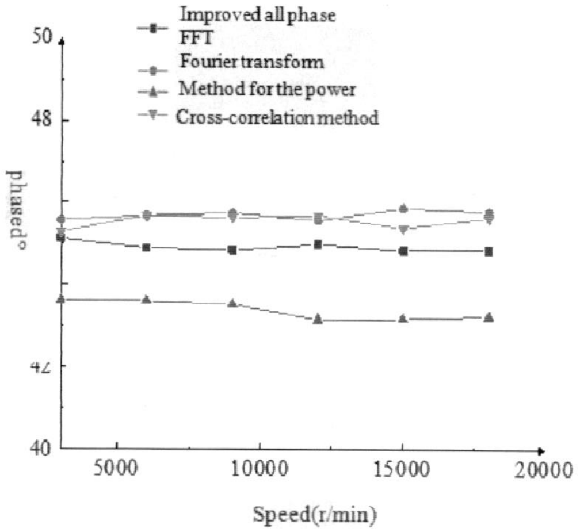

Figure 6. Phase of 40g·mm eccentric mass.

Figures 5 and 6 are comparative analysis images of the phases. It can be seen from the two phase contrast diagrams that the phases extracted by the four methods are all distributed around 45° set in the experiment.

However, the phase extraction accuracies of the four methods are completely different. The improved all-phase fast Fourier transform method is more accurate and stable in extracting the vibration phase of the main shaft.

5. Unbalance feature identification experiment

The experiment used a self-balancing electric spindle designed by gas, the eccentric mass position was set to 45°, the unbalanced mass was 50g·mm, and the rotational speed was 15000r/min. The characteristics of the vibration signal are extracted by the fast Fourier transform method and the improved all-phase fast Fourier transform method, and the dynamic balance experiment is carried out. The balance effect comparison and experimental data of the online dynamic test of the electro-spindle are shown in figure 7 and table 2.

Table 2. Spindle balancing experiment data.

Dynamic equilibrium parameter	Improved all phase FFT	The Fast Fourier transform
Initial unbalance	65 μm (45°)	60 μm (43°)
The weight added by the test	2g (90°)	2g (90°)
Weighted unbalance	23 μm (53°)	25 μm (55°)
Correction weight	3.07g (274°)	3.34g (278°)
Influence coefficient /μm/(g·mm)	21.17 (130.67°)	17.96 (124.68°)

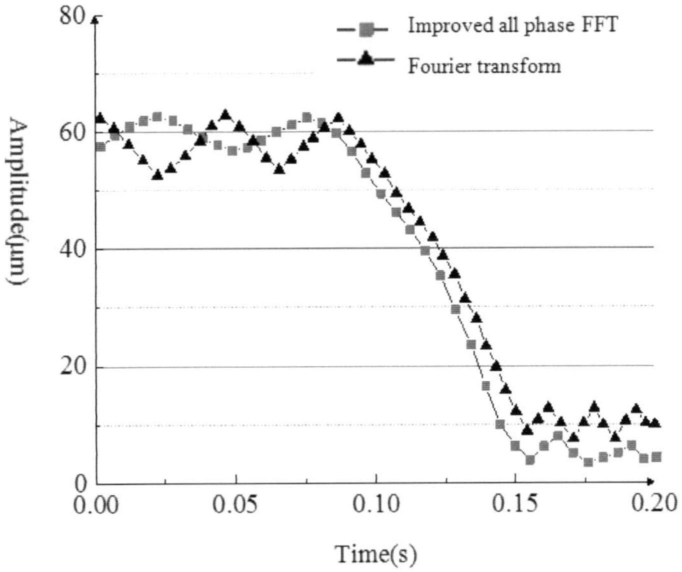

Figure 7. Comparison of effect before and after balance.

The dynamic balance experiment shows that taking the vibration signal extracted by the improved all-phase fast Fourier transform method as the input, the amplitude after dynamic balance is reduced by 98.6%, which is 10.2% lower than the calculation method built in the dynamic balance system, and the operation stability of motorized spindle is significantly improved. The results show that the improved all-phase fast Fourier transform method can effectively identify the vibration characteristics of the rotor system. Therefore, this method can be added to the dynamic balance system of the horizontal shaft of the drilling rig.

6. Conclusion

Aiming at the common vibration faults of the horizontal shaft of drilling rigs, this paper proposes a vibration feature extraction method for rotating machinery based on an improved all-phase fast Fourier transform. Taking the motorized spindle as the simulation and experimental object, the extracted spindle vibration characteristics are compared with those extracted by the Fourier transform method, the cross-power method and the cross-correlation method, respectively. The results show that the improved all-phase fast Fourier transform method can extract the main shaft vibration phase more accurately, and its stability is better than the other three methods. In the dynamic balance experiment, the dynamic balance effects of the improved all-phase fast Fourier transform method and the fast Fourier transform method are compared. The results show that the improved all-phase fast Fourier transform method is more stable, and the amplitude after dynamic balance is reduced by 98.6%, which is 10.2% lower than the calculation method built in the dynamic balance system. The conclusion proves the accuracy of the method in extracting the vibration characteristics of the rotor system, and provides a new way to solve the unbalanced vibration problem of the rotor system of the horizontal shaft of the drilling rig.

Acknowledgements

This work was supported by National Science Foundation of China: 51805337, National Science Foundation of China: 52175107, Key Laboratory of Vibration and Control of Aero-propulsion System, Ministry of Education, North-eastern University: VCAME202008 and Scientific Research Project of Liaoning Provincial Education Department: Injc202010.

References

[1] Zhou Y H, Li Y and Liu H J 2019 Feature Enhancement Method for Drilling Vibration Signals by Using Wavelet Packet Multi-band Spectral Subtraction *Strojniski Vestnik-Journal of Mechanical Engineering*. **65** 4

[2] Xu J, Zhang J J and Wang X et al. 2015 Research on unbalance vibration signal de-noising of motorized spindle *Journal of Vibroengineering*. **17** 4

[3] Xie Z J, Song B Y and Zhang Y et al. 2014 Application of an Improved Wavelet Threshold Denoising Method for Vibration Signal Processing *Advanced Materials Research*. **33**

[4] Wang Z, Zhu F L, Tu W 2018 Research on Spindle Vibration Signal Extraction by Cross-Correlation Method *Modular Machine Tool & Automatic Manufacturing Technique*. **19** 2

IOP Conf. Series: Materials Science and Engineering 1283 (2023) 012013 doi:10.1088/1757-899X/1283/1/012013

Thermal characteristics analysis of bearing in grinding system of grinding robot

Z Wang[1*], Z C Yin[1], Z N Wang[1] and C M Chen[2]

[1] School of Mechanical Engineering, Shenyang Jianzhu University, Shenyang, China
[2] General Technology Group Shenyang Machine Tool Co. LTD, China

* juven1126@163.com

Abstract. Grinding robot is used in building construction to replace manual grinding of concrete components. Bearing in grinding system under high-speed conditions has friction violently with the inner and outer ring, and generates a lot of heat. This will cause thermal failure of grinding system. In order to explore the thermal characteristics of rolling bearings, a numerical analysis model based on bearing heat transfer mechanism was established to study the influence of different working conditions on bearing temperature. The results provide a reference for reducing the thermal failure of rolling bearings in grinding system of grinding robot.

1. Introduction

The construction industry has witnessed rapid development in recent years, and the improvement of project quality is crucial for construction enterprises. Usually, concrete components and walls are polished when the construction is nearing completion. At present, the wall is mainly polished by manual hand-held grinding mechanism or sandpaper. A lot of dust can be accumulated in the process of grinding, which not only pollutes the environment, but more importantly affects the health of the staff. There is a gradual trend of construction robots replacing manual workers. The appearance of grinding robot can effectively reduce environmental pollution and improve grinding efficiency.

The grinding mechanism in the robot is the most important part to realize the function. The grinding system grinds the wall through high-speed rotation. As an important supporting part in the grinding system, the rolling bearing's reliability is very important, and the bearing thermal failure that may occur due to intense friction should be avoided.

Many researchers have done a lot of systematic research on grinding robot. Li et al. [1] studied the overall motion characteristics of the grinding robot and tested the efficiency of its grinding operation, Yao G R et al. [2] used finite element analysis software to analyse and optimize the main structure of the grinding robots, Diao S et al. [3] studied the application of three-dimensional vision systems to grinding robots. There are few researches on grinding robot bearings, and the thermal failure of bearings is very important. Many scholars have also done a lot of researches. Liu et al. [4] studied the influence of thermal effect on the fatigue life of high speed angular contact ball bearings under different working conditions, Xu et al. [5] systematically studied the mechanism and precise simulation of thermal deformation of mechanical parts, Song et al. [6] established the bearing heat generation model combined with spin heat generation to analyse the causes of bearing failure, Zhang et al. [7] simulated the contact stress and deformation of deep groove ball bearings based on Ansys, Miao et al. [8] established a three-dimensional elastic rolling contact finite element model of rolling bearings to study the dynamic characteristics of rolling bearings, Xu et al. [9] established a three-dimensional finite element model for thermal analysis of rolling bearings to analyse the influence of thermal deformation on clearance.

Based on the theorems of bearing heat generation and heat transfer and the finite element theory, the numerical simulation model of the grinding mechanism in the grinding robot is established in this paper, and the temperature field and thermal deformation of the bearing are analysed considering different working conditions. The results provide a reference for exploring the internal thermal characteristics of the bearing and a theoretical basis for avoiding thermal failure of the grinding mechanism.

2. Mathematical modelling

2.1. *Heat generation mechanism of bearing*

Bearing in grinding system under high-speed conditions has friction violently with the inner and outer ring, and generates a lot of heat. The bearing assembly in the grinding mechanism is shown in figure 1. Palmgren [10] proposed the following more accurate calculation formula (1-5) based on the measurement results of bearing friction torque.

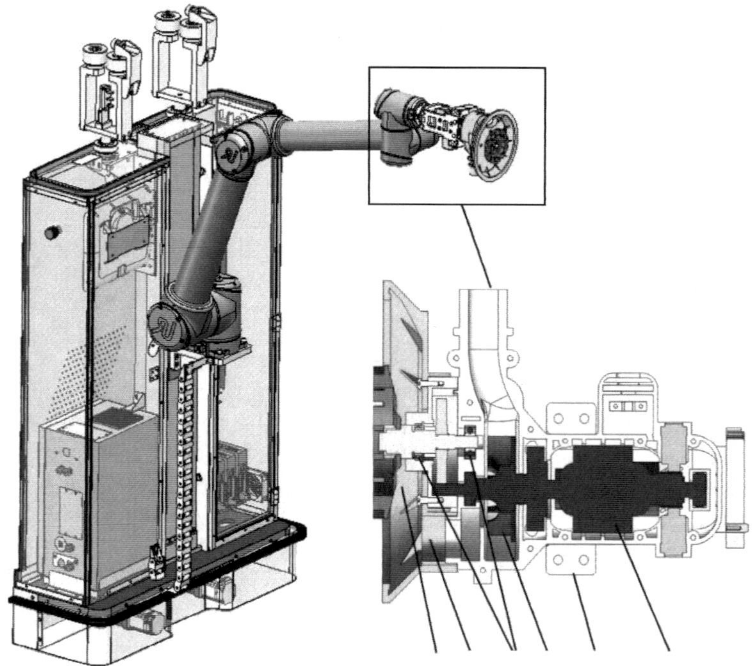

Figure 1. Grinding robot and section view of grinding system: 1. diamond grinding disc, 2. front cover, 3. bearing, 4. fan, 5. fixed mount, 6. motor.

M_r represent torque due to viscous friction of lubricating oil, it can be expressed as

$$M_r = 10^{-7} f_0 (v_0 n)^{2/3} d_m^{\ 3}, \quad v_0 n \geq 2000 \tag{1}$$

$$M_r = 160 \times 10^{-7} f_0 d_m^{\ 3}, \quad v_0 n < 2000 \tag{2}$$

Where v_0 is lubricating oil viscosity, n is speed, unit is r/min, f_0 is the coefficient related to bearing type and lubrication mode, d_m is the bearing medium diameter.

M_v represent the friction torque generated by the load, it can be expressed as

$$M_v = f_1 F_\beta d_m \tag{3}$$

Where a f_1 represents the correlation coefficient between bearing structure and load, F_β represents the magnitude and direction of the force, d_m is the diameter of the pitch circle of the bearing, unit is mm.

M_z represent the friction torque due to the spin, it can be expressed as:

$$M_z = \frac{3}{8}\mu_s QaL_{(e)}\sin\alpha \tag{4}$$

Where, μ_s is the contact friction coefficient between the rolling body and the inner and outer rings, Q is the normal contact load between the rolling body and the raceway ground, a is the roll to the long half axis of the contact ellipse, $L_{(e)}$ is the second ellipse integral of the contact surface between the angular contact ball bearing rolling body and the inner and outer rings, and α is the contact angle.

The friction torque M mainly consists of the following three components:

$$M = M_v + M_r + M_z \tag{5}$$

The calorific value caused by bearing friction is:

$$H = \pi\omega D_{pw} M / 30 \tag{6}$$

Where ω is the angular speed of the bearing, unit rad/s, D_{pw} is the diameter of the pitch circle of the ball group, unit is mm.

2.2. *The theory of heat elastic mechanics*

Thermal stress is generated in a body under external constraints and the reciprocal reduction of parts within the body. The total deformation component of the elastomer is:

$$\begin{cases} \varepsilon_x = \frac{1}{E}\left[\sigma_x - \mu\left(\sigma_y + \sigma_z\right)\right] + \alpha T \\ \varepsilon_y = \frac{1}{E}\left[\sigma_y - \mu\left(\sigma_x + \sigma_z\right)\right] + \alpha T \\ \varepsilon_z = \frac{1}{E}\left[\sigma_z - \mu\left(\sigma_y + \sigma_x\right)\right] + \alpha T \\ \gamma_{yz} = \frac{2(1+\mu)}{E}\tau_{yz} \\ \gamma_{zx} = \frac{2(1+\mu)}{E}\tau_{zx} \\ \gamma_{xy} = \frac{2(1+\mu)}{E}\tau_{xy} \end{cases} \tag{7}$$

Where ε is positive strain, γ is shearing strain, E is Tensile modulus of elasticity, μ is Poisson ratio, σ is positive stress, τ is shearing strength, α is thermal diffusivity, T is the temperature.

3. Numerical model

The bearing model used in the model is 7008c, and its geometric parameters are shown in table 1. Since the excessive fillet and chamfering of the bearing have little influence on the internal stress deformation and temperature field distribution of the bearing, they are ignored in the modeling. The 3D model is shown in figure 2.

The main structure of the bearing (inner ring, outer ring and ball) is made of GCr15Z, and the cage is made of phenolic adhesive tube. The mechanical and thermodynamic performance parameters of the two materials at room temperature are shown in table 2.

Table 1. 7008c bearing parameters.

Parameters	numeric value
bearing outside diameter/mm	68
bearing bore diameter /mm	40
Bearing width/mm	15
Rolling body diameter/mm	7.2
Number of rollers	18
Contact angle/(°)	15

Table 2. material parameter.

Materials	GCr15Z	Phenolic tape tube
Density /(kg/m^3)	7800	1800
elasticity modulus /GPa	207	2. 6
Poisson ratio	0. 30	0. 35
thermal conductivity /(W/(m·K))	30	0. 72
specific heat /(J/(kg·K))	450	1500
coefficient of thermal expansion /($°C^{-1}$)	1.15×10^{-5}	1.98×10^{-5}

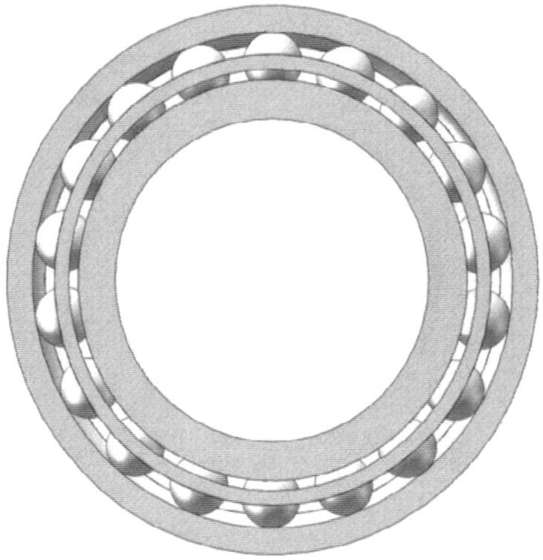

Figure 2. 3D model of bearing.

The geometric model of bearing is meshed. The unit type of thermal analysis selects the three-dimensional solid unit SOLID90 with 20 nodes. In the process of bearing thermal analysis, heat flow load and convection load need to be applied on the same surface, so the surface effect unit is introduced, and the three-dimensional surface effect unit SURF152 is established on the solid unit surface of the ball bearing and the inner and outer raceway contact area. The total number of mesh is 82382, and the mesh quality is greater than 0.8. The mesh model is shown in figure 3. For the purposes of analysis, the thermal conductivity and expansion coefficient of the material are assumed to be constant values throughout the analysis.

Figure 3. 3D mesh model.

In the finite element analysis of the thermal structure of bearings, according to the views proposed by Burton and Steph, half of the friction heat generated by bearings is transmitted to the rolling body and half to the ring [11]. The generated heat derived from equation (6) is loaded on the rolling body and the ferrule in the form of heat flux as the boundary condition. The convective heat transfer coefficient between the bearing and the air is set, and the ambient temperature is selected as 22°C. Considering the bearing seat's restriction on the outer ring of the bearing, all nodes on the outer circular surface of the outer ring are restricted in three directions: x, y and z degrees of freedom. Considering the cage's restriction on the ball, in the column coordinate system, the ball is connected to the contact points of the inner and outer rings all nodes, constrain their tangential degrees of freedom. The above model is solved based on these boundary conditions.

4. Results and discussion

Ansys workbench is used to simulate the bearing temperature field. When the rotating speed is 1000 rpm, the bearing temperature field is shown in figure 4. It can be seen that the highest bearing temperature rise occurs in the rolling body, followed by the cage, then the bearing inner ring, bearing, the outer ring temperature rise is the lowest. The inner ring has a smaller heat transfer area than the outer ring, resulting a lower temperature. Because the friction of rolling body is the most intense and the space is relatively closed, the heat dissipation is not easy. At 1000 rpm, the highest temperature of the bearing can reach 43.13°C.

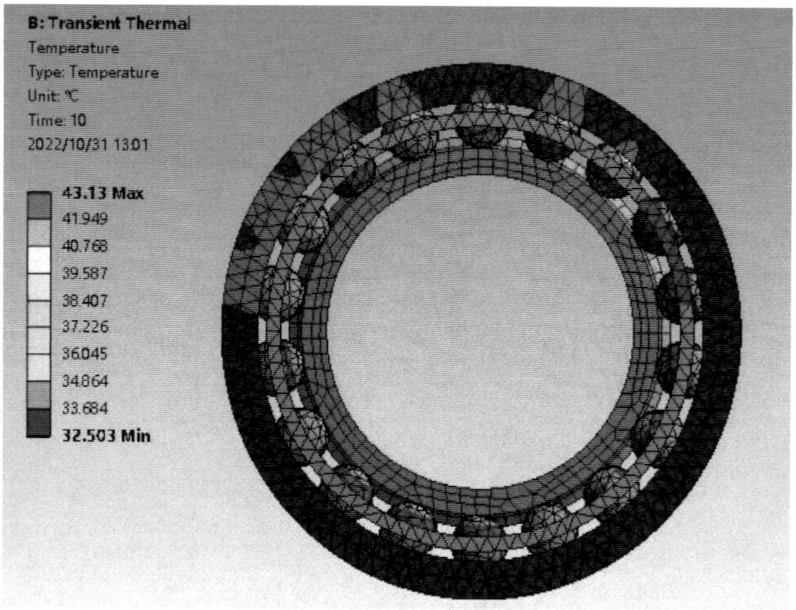

Figure 4. Bearing temperature field at 1000 rpm.

The temperature variation of bearing at different speed is analysed. The temperature variation rule of the inner and outer rings of the bearing and the rolling body as the speed changes is obtained, as shown in figure 5. It can be seen from the figure 5 that the temperature rise of the rolling is the highest and that of the outer circle is the lowest. When the speed reaches 4000 rpm, the temperature rise is the highest, reaching 55.19°C.

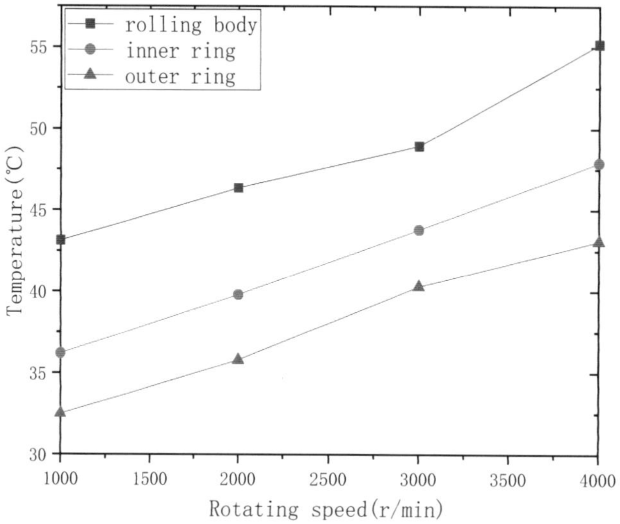

Figure 5. Point plot of bearing temperature changing with speed.

The temperature field data solved above is used as boundary conditions to analyse the thermal deformation of bearings, and a fixed constraint is added to the mounting surface of the inner and outer rings of bearings. The thermal deformation of the obtained bearing at 1000 rpm is shown in figure 6. It can be seen from the figure that the maximum rise of rolling body temperature leads to the maximum deformation. From the contact surface between the inner and outer ring and the rolling body to the fixed surface of the inner and outer ring, the deformation gradually decreases, because the heat generated by friction mainly comes from the raceway.

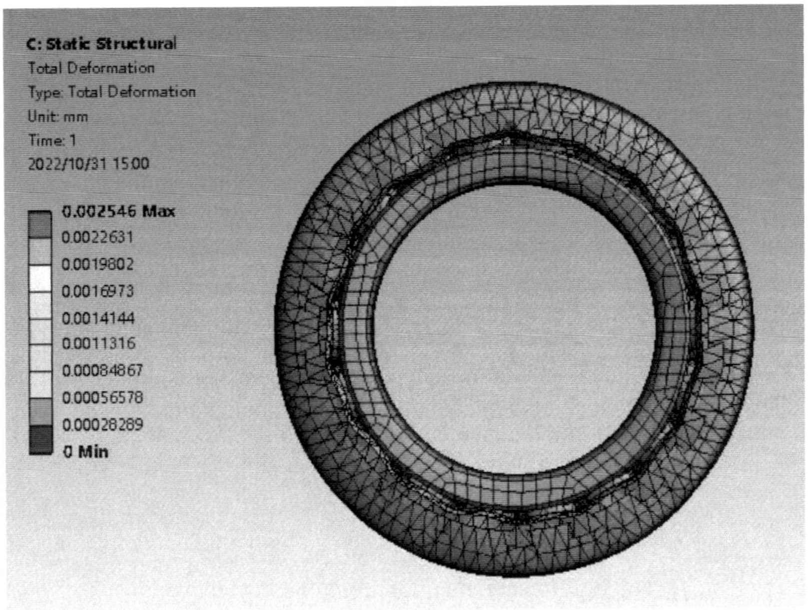

Figure 6. Bearing thermal deformation at 1000 rpm.

With the increase of the rotational speed, the thermal deformation of the bearing also increases, and the variation trend is shown in figure 7. It can be seen that when the rotational speed reaches 4000 rpm, the thermal deformation can reach up to 3.6 microns.

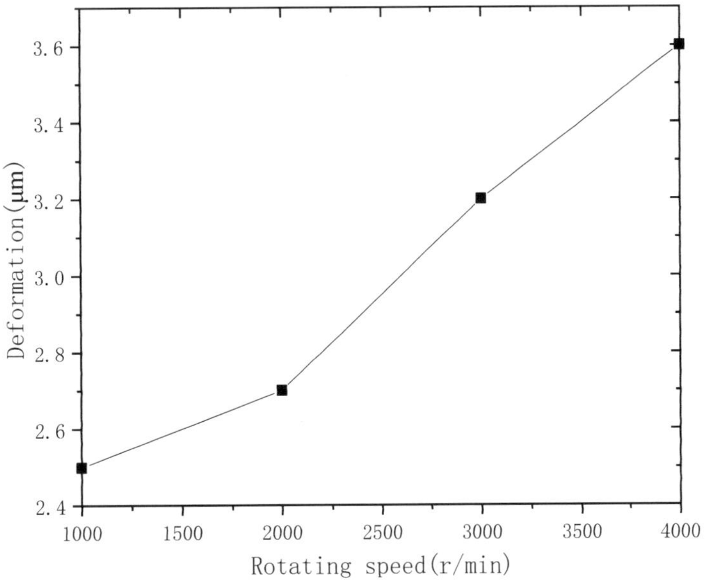

Figure 7. Point plot of bearing thermal deformation with speed change.

5. Conclusion

Based on the bearing heat generation theory and finite element theory, the finite element model of the bearing inside the grinding system of the grinding robot is established, and the influence of the speed on the bearing temperature and the influence of the thermal deformation is analyzed. The results show that:

1) when the bearing is applied to the grinding robot for grinding, the heat generated by the friction of the bearing is very large;
2) bearing friction causes temperature rise, rolling temperature rise is the highest, followed by cage, then the bearing inner ring, and the bearing outer ring has the lowest temperature rise;
3) the temperature increases with the increase of speed, and the maximum bearing temperature can reach 56.19°C at 4000 rpm;
4) the thermal deformation of the bearing increases with the increase of speed, and the maximum thermal deformation of the bearing can reach 3.6 μm at 4000 r/min.

This paper mainly studies the thermal characteristics of rolling bearings. There are many factors affecting the thermal characteristics of bearings, which are not fully considered in this paper. We will consider bearing lubrication to do further work in the future.

Acknowledgements

This work was supported by National Science Foundation of China: 52175107, 52205117, 51975388. Key Laboratory of Vibration and Control of Aero-propulsion System, Ministry of Education, Northeastern University: VCAME202008.

References

[1] Li Y F 2021 Key Technology Research on Climbing Scaffold Type Robot for External Metope Polishing. Shanghai university
[2] Yao G R 2020 Structure Design and Finite Element Analysis of Grinding Robot. Shenyang university of technology
[3] Diao S, Chen X, Luo J 2018 Development and experimental evaluation of a 3D vision system for grinding robot. Sensors
[4] Liu Xiaowei, Wang Wei, Wang Qingli 2015 Study on Temperature Field of High Speed Angular Contact Ball Bearings Based on ANSYS. Journal of combination machine tools and automatic processing technology, (3): 13-15 + 20
[5] Xu Zhishang 2015 The mechanism and simulation research on thermal deformation of mechanical parts. Hefei University of Technology
[6] Song Nan 2017 Analysis on Temperature Field and Thermo-Stress Coupling of High Speed Angular Contact Ball Bearing with Spin Heat Generation. Jilin University
[7] Zhang Fuxing, Zheng Yuan, Wang Qing, Li Jie, Xu Suhong, Xin Huarong, Wu Zhong 2012 Contact Stress FEM analysis of Deep Groove Ball Bearing Based on ANSYS Workbench. Mechanical design and manufacturing, (10) : 222-224
[8] Miao Yingyun, Chen Xiaoan 2011 Elastic contact dynamics of rolling bearing based on finite element method. Journal of Mechanical Strength,33(05):708-713
[9] Xu Jianning, QU Wentao, Zhao Ning 2006 Analysis on Temperature Field and Thermal Deformation of Rolling Bearings. Bearing, (05):1-3
[10] Palmgren A 1959 Ball and Roller Bearing Engineering. Philadelphia: Burbank,
[11] An Jiaqi 2020 Thermal Structure Coupling Analysis of Angular Contact Ball Bearings. Shenyang University of Chemical Technology.

Reuse of mixed masonry wastes in plastering mortars

A Zaharie[1], M L Țințișan[1], AC Siomin[1], D L Manea[1] and M L Pleșa[1]

[1] Technical University of Cluj-Napoca, Faculty of Civil Engineering, Cluj-Napoca, Romania

* annazaharie@yahoo.com

Abstract. According to Eurostat statistics, the volume of waste collected is constantly increasing, the amount of waste produced in Europe in 2020 reaching 2151 million tons. In the European Union, the most waste-producing activity is construction and demolition, accounting for 37.1% of the total waste produced in 2020. This type of waste has a high occupancy rate. This article proposes the reuse of mixed masonry waste in the production of plaster mortars. The maximum amount of mixed masonry waste used in the experiment was 35%, partially replacing the aggregate component. The physical and mechanical characteristics that have been measured in the experimental program may be divided into two categories: characteristics of mortars in fresh state (apparent density, consistency) and in hardened state (bending strength, compressive strength and adhesion to the support layer). In order to highlight the influence of the addition of waste on the composition and on the evolution of the mechanical properties of the studied mortars, the experimental program was distributed over 3, 7 and 28 days. The results showed an increase in the mechanical strength of mortars containing mixed masonry waste compared to standard mortar. Mixed masonry waste can be a substitute for raw materials in the production of new types of plaster mortars and contributes to the reduction of the resulting waste by recycling and reuse. The experimental program will be continued with microscopic investigations using modern investigation methods, to highlight the behaviour of mortars with masonry waste, and to identify the correct proportion of waste that can be used in mortar networks so as not to affect the characteristics required by the acting norms.

1. Introduction

Waste reuse aims to reduce the negative impact on the environment, the amount of generated waste and the space needed for waste storage, to improve the quality of life and to make the consumption of natural resources more efficient.

According to Eurostat statistics, the amount of waste generated in 2020 in the European Union is predominantly from construction and demolition activities followed by mining and quarrying waste [1]. Reducing the amount of generated waste can be achieved by maximizing the reuse process for the finished product or recycling the waste in new materials or finished products.

There are many research studies focusing on recycling of different types of waste in the construction industry, as recycled aggregates produced from construction and demolition waste to obtain new building materials, protect the environment by using fewer natural resources and reduce overall pollution.

According to the bibliometric analysis in the scientific database Scopus, there are currently 35 scientific publications with "masonry waste" as a keyword. The year 2016 featured the most publications on this topic, the year 2022 registering the last 3 articles published on this subject. Engineering and Materials Science are the major area of interest in the reuse of masonry waste [2].

The reuse of masonry waste has been studied in order to obtain cementitious materials in which mixed masonry waste partially replaces the cement or aggregate components [3-6]. Masonry construction and demolition activities produce a large quantity of brick or concrete masonry waste.

The aim of the present paper is to identify a method to use mixed masonry waste as a replacement for aggregates in the composition of plastering mortars, in as high percentage as possible (10%, 25% and 35%). In Romania, the reuse of masonry waste is deficient [2]. The mixed masonry waste used in this study is in a large proportion composed of ceramic brick, a predominant material used in Romania since the 1900s in the construction of residential buildings.

2. Materials and recipes

In the experimental program, the aggregate-type solid component of the standard CS IV mortar recipe was replaced with mixed masonry waste. The standard CS IV(MS) plaster mortar recipe for 1 m^3 prescribes the following quantities of materials: 385 kg Portland cement 52,5R, 1541 kg sand sort 0-4 mm and water.

The four recipes carried out in the experimental program are: CSIV standard plaster mortar (MS), plaster mortar with 10% mixed masonry waste (10% MMW), plaster mortar with 25% mixed masonry waste (25% MMW), plaster mortar with 35% mixed masonry waste (35% MMW) (table 1).

In figure 1 are presented the mixed masonry wastes (which was crushed) and the sand used in the experimental program which were sorted by grain size.

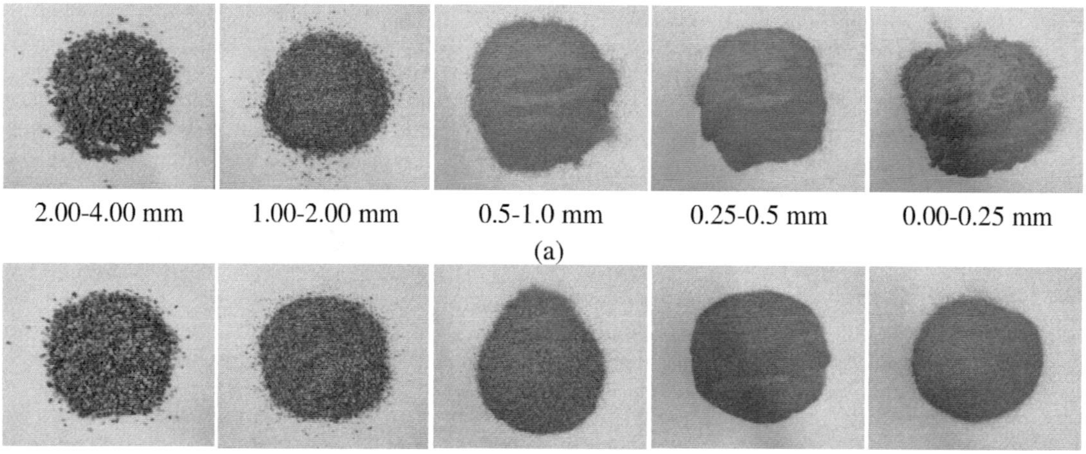

2.00-4.00 mm 1.00-2.00 mm 0.5-1.0 mm 0.25-0.5 mm 0.00-0.25 mm

(a)

2.00-4.00 mm 1.00-2.00 mm 0.5-1.00 mm 0.5-0.25 mm 0.00-0.25 mm

(b)

Figure 1. a) Sort of mixed masonry waste; b) sorts of sand.

Each sand sort has been partially replaced with mixed masonry as shown in table 1.

To obtain a specific consistency of plaster mortar, the quantity of water varies from recipe to recipe, depending on the amount of waste added. The water/cement ratio varies between 0.52 to 0.69; for the mortars with mixed masonry waste the ratio is lower than that of standard mortar. This variation is caused by the fact that the mixed masonry waste contains an uncontrolled amount of solid brick, plastering dust and paint. The values indicate that the composition of proposed mortars with wastes is more compact than standard mortar.

Table 1. Recipes of experimental program.

	Mortar types	MS	MMW 10%	MMW 25%	MMW 35%
	Cement [kg/m^3]	385	385	385	385
Fine aggregates [kg]	0÷0.25 mm	77.5	69.75	58.13	50.38
	0.25÷0.50 mm	232.5	209.25	175.4	151.13
	0.50÷1 mm	378.5	340.65	283.9	246
	1÷2 mm	232.5	209.25	175.4	151.13
	2÷4 mm	620	558	465	403
Masonry waste [kg]	0÷0.25 mm		7.75	19.38	27.13
	0.25÷0.50 mm		23.25	58.13	81.38
	0.50÷1 mm		37.85	94.63	132.5
	1÷2 mm		23.25	58.13	81.38
	2÷4 mm		62	155	217
	w/c ratio	0.69	0.52	0.61	0.63

3. Methods

The mortar-type building material used as a binder of two solid materials or as a waterproofing, finishing and lining of walls and floors must comply with the required European standards. The methods used in the experimental program are aimed at determining the physical and mechanical characteristics of mortars made with mixed masonry waste. The physical and mechanical characteristics that have been determined are as follows: apparent density, mortar consistency, compressive strength, tensile strength and adhesion to the substrate.

Apparent density was measured on the ecological mortar in fresh and hardened state. Apparent density is an important characteristic of building materials and is closely related to other parameters such as hardness, conductivity and material compatibility. According to SR EN 1015-6 2001/A1:2007 [7] and SR EN 1015-10: 2002 / A1: 2007 [8], the apparent density of mortar in fresh and hardened state was determined.

The consistency of the mortar is directly influenced by the dosage of the component materials in the mortar recipe. Consistency is determined by the spreading mass method on fresh mortar and is measured in mm. The consistency of the material is determined according to the intended use of the material according to SR EN 1015-3:2001 [9] by measuring the diameter of the mortar mass being spread.

Regarding the mechanical characteristics of the mixed masonry mortar specimens the flexural and compressive strength of the hardened mortars was determined according to SR EN 1015-11:2020 [10].

The determination of the adhesion to the substrate was carried out by tensile test after hardening of the mortar on the ceramic substrate according SR EN 1052-3 [11].

4. Experimental program

In the experimental program the physical and mechanical characteristics of the proposed recipes were determined both in fresh and hardened state. Three mortar specimens were made and tested for each of the proposed recipes. According to the acting norm, the average value from three test specimens will be used.

Figure 2 shows the samples of mortars with 25% mixed masonry waste proposed for testing at 28 days.

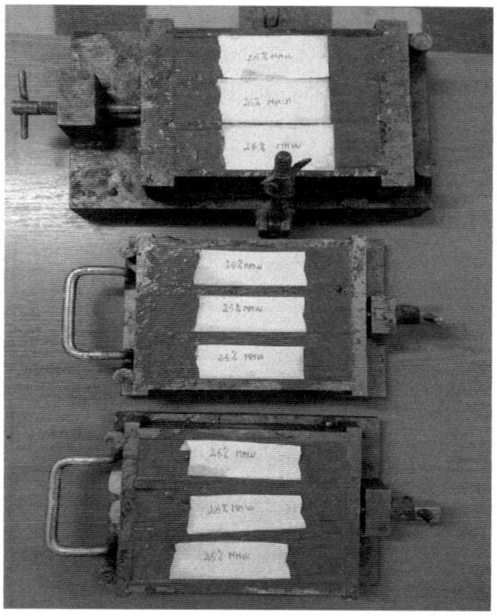

Figure 2. Samples of mortars with 25% mixed masonry waste.

4.1. Physical characteristics
The measured physical characteristics of the mortars are the apparent density in fresh and hardened state and the consistency with spreading table.

4.1.1 Apparent density of fresh mortars. The highest value is that of the standard mortar and all the values for the mortars with brick wastes are lower, which indicates that by using wastes as a replacement for aggregates the apparent density decreases (the specimens are lighter).

Table 2. Apparent density on fresh mortars.

Mortar types	MS	MMW 10%	MMW 25%	MMW 35%
Fresh mortar apparent density [kg/m^3]	2242	2050	2185.5	2125

According to table 2, the use of mixed masonry waste generates a decrease in the apparent density of fresh mortars.

4.1.2. Apparent density of hardened mortars. By partially replacing the sand with mixed masonry waste, a decrease in apparent density is observed, proportional to the amount of added waste. The apparent density of hardened mortars is between 1984-2210 kg/m^3. The mortar with the highest value is the MMW 10% recipe.

Table 3. Apparent density of hardened mortars.

	Mortar age					
	3 days		7 days		28 days	
Mortar Types	Weight [kg]	Apparent density [kg/m^3]	Weight [kg]	Apparent density [kg/m^3]	Weight [kg]	Apparent density [kg/m^3]
MS	0.575	2246	0.563	2200	0.567	2210
WWM 10%	0.541	2113	0.545	2129	0.535	2090
WWM 25%	0.540	2110	0.530	2070	0.530	2070
WWM 35 %	0.522	2039	0.512	2000	0.508	1984

4.1.3. Mortar Consistency. The consistency is measured for the mortars in fresh state and the results are presented in the table 4.

Table 4. Mortar consistency values.

Mortar type	MS	MMW 10%	MMW 25%	MMW 35%
Average value of the consistency [mm]	182.5	183.0	199.0	200.0

According to the obtained results, the addition of mixed masonry waste generates an increase in the consistency values of fresh mortars. The mortar with the lowest value is standard mortar, which indicates that the components of the mortar are well bonded. According to the obtained consistency values, the studied mortars can be used in plastering.

4.2. Mechanical characteristics
To analyse the impact of the added waste on the mortars in hardened state, the samples were tested at 3, 7 and 28 days.

4.2.1. Bending tensile strength. In order to identify the influence of the addition of masonry waste on the mechanical characteristics of mortars, the experimental program includes measurements at the ages of 3, 7 and 28 days, thus highlighting the evolution of the mechanical properties with time. The tensile strength was determined at hose exact ages.

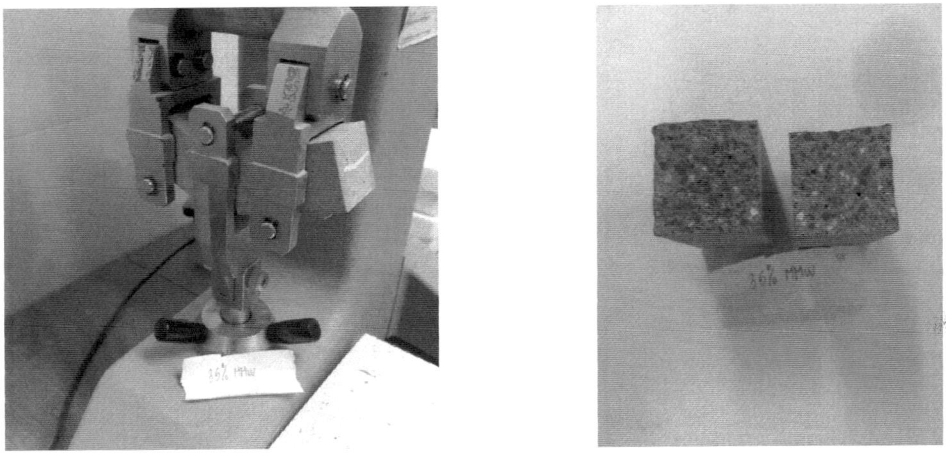

Figure 3. Mortar sample test of bending tensile strength with 35% of mixed masonry waste.

Table 5. Values of Bending tensile Strength [N/mm^2].

Mortar type	MS	MMW 10%	MMW 25%	MMW 35%
Mortar age				
3 days	4.47	3.73	4.60	4.40
7 days	5.22	5.91	4.88	5.09
28 days	6.08	6.20	6.22	5.74

The tensile strength increases with time for all the studied recipes.

According to table 5 and figure 4, all recipes with mixed masonry waste have higher values than the standard mortar. The mortar with the highest value of tensile strength is the mortar in which the aggregates have been replaced in a proportion of 25%.

The evolution of the bending tensile strength in time is presented in Figure 4.

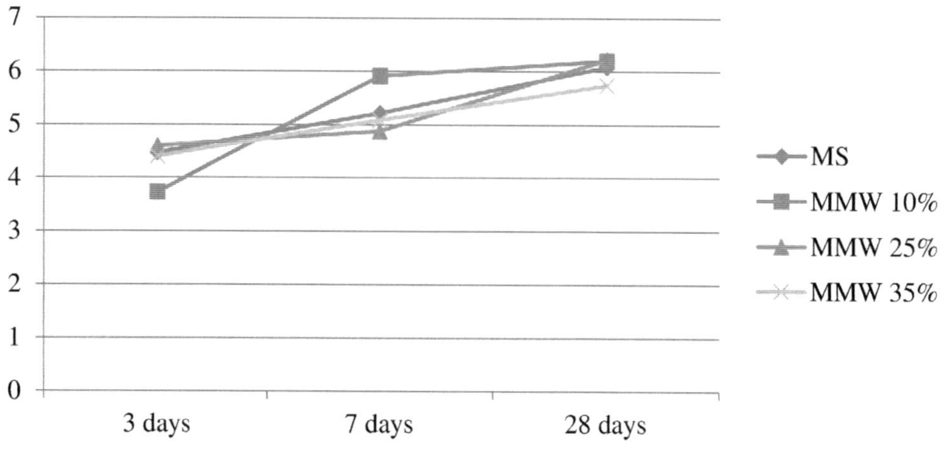

Figure 4. Graphic of Bending Tensile Strength at 3, 7 and 28 days.

4.2.2. Compressive strength. As in the case of consistency, the mortars with 25% and 35% waste have almost the same values. All the mixed masonry waste mortars exhibit lower compressive strength than the standard mortar.

According to figure 3, the values for the mortars with mixed masonry waste are close to those of the standard mortar.

Table 6. Average values of compressive strength [N/mm^2].

Mortar type	**MS**	**MMW 10%**	**MMW 25%**	**MMW 35%**
Mortar age				
3 days	34.86	33	24.01	24.02
7 days	36.71	35.98	31.49	30.85
28 days	39.11	38.91	36.47	36.93

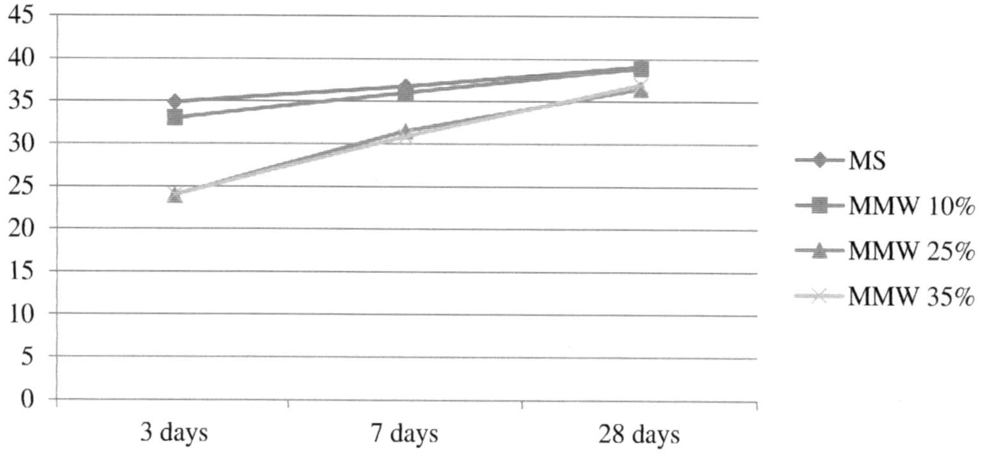

Figure 5. Graphic of Compressive Strength at 3, 7 and 28 days.

4.2.3. Adhesion of the support layer. For the adhesion to the support layer determination, the samples were kept in the wet air box until testing, at 28 days. The perpendicular force applied to the hardened mortar surface gives the shear strength of the surface. The support layer is pressed solid brick.

Figure 6. Samples of fresh and hardened mortar state with 25% mixed masonry waste

Table 7 presents the results of the adhesion to the support layer tests, for all of the studied mortar types (MS, MMW 10%, MMW 25%, MMW 35%).

Table 7. Adhesion to the support layer at 28 days.

Mortar type	MS	MMW 10%	MMW 25%	MMW 35%
Adhesion to the support layer [N/mm^2]	0.035	0.065	0.050	0.040

The adhesion to the support layer with the highest value of over 0.065 N/mm^2, was found for MMW 10%. All the results obtained on the mortars with recycled aggregates have higher values than standard mortar, which shows that the waste introduced in the recipe does not negatively influence the adhesion to the support layer. The lowest value was obtained for the standard mortar (MS).

5. Conclusions

The amount of waste resulting from construction and demolition can be reduced by using the mixed masonry waste to obtain new mortar recipes. The mixed masonry waste can be collected and reused directly on site.

Analyzing the quantities of mixed masonry waste used in the experimental study, by replacing only 10% of the component aggregates, 154 kg of waste can be recycled for each cubic meter of mortar. For a partial replacement of 35 % of the aggregates, 539 kg of waste can be reused for 1 m^3 of mortar. By reusing even the smallest amount of waste, a significant reduction in environmental pollution may be obtained.

According to the results obtained in the experimental program, the characteristics of all the studied mixed masonry mortars are within the limits of the relevant norms. The recipe that has obtained the closest characteristics to standard mortar is the one where 10% of the sand has been replaced with mixed masonry waste.

As regards to the characteristics determined on the fresh sample, the apparent density and consistency values are lower than those of the standard mortar. The lowest apparent density value was reached for the MMW 10% sample, with a difference of 8.56% compared to the standard sample. The mortar with the lowest and closet consistency value is the MMW 10% sample.

For the characteristics of the mortars in hardened state, the addition of mixed masonry waste generates an improvement for MMW 10% and MMW 25%, the highest increase in values being of 2.30% for MMW 10%. In terms of compressive strength, the mortars tested at 28 days present a small decrease in values. The biggest difference is 5.57%, compared to standard mortar and the closest value is that for MMW 10%.

The mortars with mixed masonry waste have a better adhesion to the substrate than the etalon mortar. The highest value of mortar adhesion is for MMW 10%.

A moderate variation of values is observed in the results, which is caused by the components materials of this mixed waste. Therefore, a molecular analysis and an RMN investigation can provide us information to optimize the recipes studied in this experiment.

Acknowledgments

This paper was financially supported by the Project "Network of excellence in applied research and innovation for doctoral and postdoctoral programs / InoHubDoc", project co-funded by the European Social Fund financing agreement no. POCU/993/6/13/153437.

References

[1] "*Eurostat Statistics Explained*," 13 01 2023. [Online]. Available: https://ec.europa.eu/eurostat/statistics-explained/index.php?title=Waste_statistics#Waste_generation_excluding_major_mineral_waste. [Accessed 24 03 2023]
[2] "*Analyze search results,*" [Online]. Available: https://www.scopus.com/term/analyzer.uri?sid=809256149a00e078564b1aded1cb9b81&origin=resultslist&src=s&s=TITLE-ABS-KEY%28%22masonry+waste%22%29&sort=plf-f&sdt=b&sot=b&sl=30&count=35&analyzeResults=Analyze+results&txGid=b3db23912f246ea17a0a872de64bcd24. [Accessed 24 03 2023]
[3] Maciá M, E, Castillo A, Martinez I, Rubino F J 2002 *High-Temperature Residual Compressive Strength in Concretes Bearing Construction and Demolition Waste (CDW): An Experimental Study*, Iranian Journal of Science and Technology - Transactions of Civil Engineering, vol. 46, no. 6, pp. 4303-4312
[4] Salazar R R, Valencia Saavedra W, Gutierrez R M, 2022 *Reuse of Powders and Recycled Aggregates from Mixed Construction and Demolition Waste in Alkali-Activated Materials and Precast Concrete Units*, Switzerland- Sustainability, vol. 14, no. 15, p. 985
[5] Salazar R R, Valencia Saavedra W, Gutierrez R M, *2022 Recycling of concrete, ceramic, and masonry waste via alkaline activation: Obtaining and characterization of hybrid cements*, Cali-Colombia-*Journal of Building Engineering,* vol. 46
[6] Sruthy S, Sriram Kumar G, Barrath S S, Ashim Suriya R, 2016 *Use of recycled brick masonry waste in concrete*, India-International Journal of Earth Sciences and Engineering, vol. 9, no. 2, pp. 541-544
[7] SR EN 1015-6:2001/A1:2007 Methods of testing masonry mortar. Part 6: Determination of apparent densisty of fresh mortar
[8] SR EN 1015-10:2002/A1:2007 Methods of testing masonry mortar. Part 10: Determination of apparent density of hardened mortar
[9] SR EN 1015-3:2001 Methods of testing masonry mortar. Part 3: Determination of consistency of fresh mortar (with spreading table)
[10] SR EN 1015-11:2020 Methods of testing masonry mortar. Part 11: Determination of bending and compressive strength of hardened mortar
[11] SR EN 1052-3:2003 Masonry test methods. Part 3: Determination of initial tensile strength

AUTHOR INDEX

Adam, L 1
Bărbuţă, M 63, 89
Bianco, A 19
Bianco, L 19
Chen, C M 113
Ćipranić, I 71
Cojocaru, A 28
Constantinescu, S 37
Diaconu, A C 63
Diaconu, L I 63
Du, S Y 106
Ezhov, V 49
Fechete, R 79
Fedyanin, D 49
Ioniţoaiei, P 55
Isopescu, D N 1, 28
Janczak, N 100
Jevrić, M 71
Lepadatu, D 1
Lu, C Q 106
Luca, B I 63
Ma, J 106
Manea, D L 79, 121
Maxineasa, S G 28
Medojević, A 71
Ostojić, M 71
Pakhomov, V 49
Panaite, V 89
Panţiru, A 63
Plesa, L M 79
Pleşa, M L 121
Rotaru, A 55
Rujanu, M 63
Saikov, I 49
Saitis, C 79
Saitis, L 79
Scripcă, S 89
Sekulić, G 71
Siomin, AC 121
Tepes-Onea, F 100
Timu, A 63
Ţinţişan, M L 121
Wang, Z N 113
Wang, Z 106, 113
Yin, Z C 113
Zaharie, A 121